UNDER NORTHERN SEAS

# UNDER NORTHERN SEAS

Photographed and written
by
LINDA PITKIN

PUBLISHED BY
SALAMANDER BOOKS LIMITED
LONDON

A SALAMANDER BOOK

Published by Salamander Books Limited
129–137 York Way
London N7 9LG
United Kingdom

9 8 7 6 5 4 3 2 1



Produced by Jill Hollis and Ian Cameron
for Cameron Books, PO Box 1, Moffat,
Dumfriesshire DG10 9SU, Scotland



Colour reproduction by Alfacolor, Verona
Chlorine-free paper made by Leikam, Austria
Printed in Italy by Artegrafica, Verona

ISBN 0 86101 973 3

A CIP record for this book is available from the British Library

Special thanks are due to the following people:

Members of the British Society of Underwater Photographers, who have given me advice, encouragement and inspiration ever since I took up underwater photography. My colleagues at The Natural History Museum, London, particularly the specialists in the Department of Zoology and the staff of the Libraries, including Geoff Boxshall, Paul Clark, Carol Gokce, Paul Cornelius, Oliver Crimmen (and other members of the Fish Section), David George and John Taylor have given support in the course of the research for this book. In addition, I am indebted to many published works consulted for reference, some of which are detailed in the bibliography. I am grateful to Jill Hollis and Ian Cameron (Cameron Books) for many helpful suggestions in the preparation of the book.

The following, and others, have organised diving for me: Nils Aukan (Norway), Bill Bunting (*Nirvana*, England); Dave Burton (*Jean de la Lune*, Scotland); Tom Dickie, Mike Glover and Ken Watterson (Isle of Man); EXUL diving club (England); Cyril Gubby (*Bubbles II* and *Triple Brande*, England); Ivor Johnson (*Osborne Bay*, Wales); John and Shaunne Shaw (Aquaserve, Lundy, England); Olav Magne Strømsholm and the staff of Strømsholmen Sjøsportsenter (Norway); Thornton Heath BSAC (Great Britain).

Above all, I wish to thank my husband, Brian, whose help, in and out of the water, has been an invaluable contribution to this book.

*Endpapers*
*Sea anemones,* Sagartia elegans, *covering the side of a wreck.*

*Half title*
*A feather duster among red seaweed; the two whorls of tentacles (the gills) are all that can be seen of the tube worm* Bispira volutacornis.

*Title spread*
*Small fish milling above plumose anemones (*Metridium senile*) in a sheltered section of a wreck at a depth of about 20 metres.*

*Contents*
*A dahlia anemone (*Urticina felina*), with (bottom left) a glimpse of the sponge-like soft coral, dead-man's fingers (*Alcyonium digitatum*), among other varied life on a rock face.*

*This page*
*A tiny hermit crab in a shell, about 2 cm long, patrols a rose-tinted landscape of rock encrusted with coralline algae.*

*For my husband, Brian.*

# CONTENTS

# INTRODUCTION

Beneath the sea's surface, brightly patterned fish swim over a colourful carpet of living animals – soft corals, sea anemones, starfish and a profusion of others, equally exotic in appearance, cover the sea bed. A tropical coral reef, you might imagine, but no, this is northern Europe, in the cool seas of the North Atlantic.

Unless you have had the opportunity to dive there and see for yourself, the sight of uninviting, dark grey water and waves relentlessly pounding the rocks makes it hard to picture what might lie below the surface. Some of the creatures are animals of the seashore, like beadlet anemones and dog whelks, which will be familiar from childhood paddling expeditions and explorations of rock pools; others, such as starfish and seashells, are often seen cast up by the high tide along the strand line. But once out of their natural surroundings, few of them retain much of the splendour they showed in life, even before they have begun to decompose. Out of water, soft-bodied animals such as jellyfish and sea anemones lose their shape immediately, and the colours of most animals fade rapidly. Even animals like crabs, lobsters and scallops – well known as seafoods – have a very different appearance in life. The row of beady eyes of the scallop, peeping out of its shell through a fringe of tentacles, the surprisingly varied facial features in species of flatfish, the plaice with its thick lips and the sole with emerald eyes – these are all ample reward for anyone venturing underwater. Who, on seeing monkfish displayed headless and skinless at the fishmonger, could visualise the startling grimace of the living anglerfish?

Diving in the sea offers the pleasure of observing marine animals face to face. If you are careful and patient, so as not to disturb them unduly, they will tolerate the intrusion, and you will be treated to a grandstand view of the dramas of everyday life underwater. Watching these activities is a captivating experience, whether you are observing slow-moving molluscs such as sea slugs or shimmering shoals of streamlined fish like pollack. Perky tompot blennies peep inquisitively out of holes in rocks and wrecks, and hovering cuttlefish flash their changing colours in a quicksilver response to a sudden encounter.

This hardly begins to describe the richness of marine life in northern seas. A wealth of fascinating fish and other animals includes such highlights as the male cuckoo wrasse, brilliantly

*Left. Tall fronds of kelp and other seaweeds top the rocks in water a few metres deep.*

*Facing page. Small fish swimming over fronds of kelp in the sunlit shallows.*

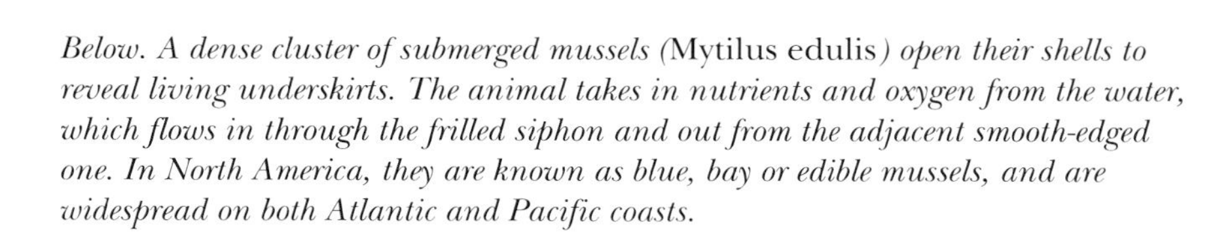

*Below. A dense cluster of submerged mussels (*Mytilus edulis*) open their shells to reveal living underskirts. The animal takes in nutrients and oxygen from the water, which flows in through the frilled siphon and out from the adjacent smooth-edged one. In North America, they are known as blue, bay or edible mussels, and are widespread on both Atlantic and Pacific coasts.*

arrayed in his courting colours, vivid red cushion stars, yellow sunset corals, and delicate sea fans in pastel shades. For me, it is often the smallest animals that have the greatest appeal: some of the nudibranch sea slugs, for instance, are exquisite in their jewel-like details. Taking the time to examine a patch of the sea bed closely is well worth the effort; there is much to see, even when a fog of sediment particles prevents more distant viewing.

The underwater scenery of northern seas is diverse: the bedrock forms low, shelving ridges in some places, dramatic pinnacles and cliffs in others. Swimming along a sheer, towering rock face is an experience that takes some beating, but there are other, equally exciting and entirely different landscapes to savour in the world beneath the waves. Few places can be more beguilingly beautiful than the tranquil inlet I visited on a Norwegian skerry, as the afternoon sunlight filtered through the green fronds of the tall kelp that formed a canopy over a gully. Swimming into this passage, you look around and find that the quiet backwater actually teems with small fish, sea slugs, starfish and many other animals, jostling one another for territory. As you leave the shade of the kelp forest, with its mysterious passages running among the almost impenetrable jungle of stems, the open vista of an adjacent sandy plain offers yet another contrast.

When I visit a site underwater for the first time, I am invariably filled with anticipation: peer into a small cave, and you may see a lobster, swim into a large cavern and you might confront a seal. Returning to a familiar place, on the other hand, brings with it both the pleasure of renewing one's acquaintance with old friends, and also unexpected surprises; day by day and from season to season, the continuous change in the sea means that even a well-known site is likely to come up with fresh sights and experiences.

The various chapters in this book cover a range of contrasting environments under northern seas, featuring a sample of the amazingly varied inhabitants, from simple sponges to seals, tiny cowries to the enormous basking shark. All the animals and plants illustrated within this book are accessible to a diver with nothing out of the ordinary in the way of equipment. Some of them live exclusively in a particular habitat, while others are more eclectic in their choice of home. Many flatfish, including plaice and brill, for instance, normally live on soft ground such as sand or gravel, while topknots select rock surfaces, but lemon sole may well be found on both of these very different types of sea bed. Other habitats have an obvious overlap and attract similar denizens: the hard structures of wrecks, for example, offer alternative homes to many rock dwellers. Each chapter illustrates a theme with typical examples, but neither habitats nor their occupants exist in isolation – their interrelationships are seemingly infinite.

The bizarre beauty of marine animals, so different from terrestrial forms of life, enthrals observers and prompts many questions. Why does an anglerfish have a beard-like frill round its head and body? Why does a tube worm have a plume of

tentacles? Study the animals' natural behaviour, and the answer is often revealed. The various characteristics of a species are adaptations that enable it to survive in its particular way of life. Investigations by scientists and dedicated amateurs have advanced our knowledge of how form relates to function, though much is still to be discovered.

*Above. A dense shoal of bib (*Trisopterus luscus*), a member of the cod family.*

*Facing page. Covered in fragments of sponge plucked from its surroundings, a scorpion spider crab (*Inachus*) perches between sea anemones (*Sagartia elegans*) on a steep wall.*

*Left. Coating a rock face, yet still reproducing, tightly grouped sea squirts are budding here to form many more tiny individuals.*

Northern seas, like marine environments worldwide, are seriously threatened by industrial pollution. This is equally deplorable whether it results from the deliberate discharge of toxic wastes inland or at sea, from the seepage of pesticides and fertiliser from agricultural land into our rivers and thence to the ocean, or from oil or other accidental spillages. Most of the blame for these depredations tends to be laid at the feet of industrialists and governments. However, as the consumers who take for granted as our right a way of life that implicitly endorses these practices, we cannot absolve ourselves of responsibility. It is as a direct result of our demands that increasing pressure is placed on the natural, self-regulating systems that maintain the balance in the sea's environment, where the fortunes of predator and prey are linked in complex food webs.

Current legislation does not go far enough to stem the damage to our seas. A few vulnerable sites that are particularly rich in marine life have been designated protected areas, but even where this type of measure is effective, the benefits to the sea at large are limited. Tides, currents and winds are no respecters of boundaries drawn by humans. The resources of the sea will be sustained for the benefit of future generations – both in the sea and on land – only if we use them wisely now, enjoying but not destroying them, and putting an end to fishing practices that extract more stocks than can naturally be replenished.

# CHAPTER ONE

# BETWEEN THE TIDES

Life at the water's edge is fraught with dangers and difficulties. The constant battering of waves breaking on the shore, the ebb and flow of the tide, and the sometimes huge variations in temperature and salinity all add up to a particularly stressful environment. Yet a wide range of animals and seaweeds are adapted to survive here and are commonly seen from the strand line (high-tide mark) to the shallows of a few metres' depth, and in rock pools between. Many animals anchor themselves firmly to rock to avoid being swept away by strong waves. Barnacles survive encased in a fortress of plates, and thick shells enclose molluscs such as limpets, preventing them from being crushed and protecting them from desiccation when they are left high and dry above the water line.

Sunlight filters through the shallow water in abundance – a major advantage for plants that need the sun's energy for photosynthesis. The seaweeds here range from tough, leathery brown wracks and kelps to more flimsy green species, such as sea lettuce (*Ulva lactuca*).

Seaweeds and the fine growth of algae on rocks provide food for molluscs, among them limpets and certain top shells, while green sea slugs (*Elysia viridis*) and the larger sea hares (*Aplysia punctata*) graze on subtidal algae. Other animals use seaweeds for cover or, like purse sponges (*Scypha*), attach themselves to it. Some hydroids, for example, grow over various brown algae indiscriminately, while others show a particular preference, like the feathery *Aglaophenia pluma* for pod weed (*Halidrys siliquosa*).

One of the most familiar animals living in the shallow water near the shore – the sublittoral zone – is the common starfish (*Asterias rubens*). Though variable in colour, it is typically orange with small white spines over its body and rather plump arms. It is most likely to be seen on sand or around mussel beds, which, along with scallops, provide an important part of its diet.

Among other inhabitants of the shallows are the candy-stripe flatworm and the sea lemon (*Archidoris pseudoargus*). The sea lemon is one of the best-known northern European sea slugs, often found from the lower shore down, and, at up to 12 cm in length, one of the largest species. Its mottled warty body is generally yellowish (hence its common name) and has a conspicuous ring of bushy gills on the lower back, which are rapidly withdrawn when the animal is disturbed. The sea lemon feeds on sponges, particularly the breadcrumb (*Halichondria panicea*), which forms green patches on rock surfaces.

Only a few flatworms live around the coasts of northern Europe, but one of these, the candy-stripe flatworm (*Prostheceraeus vittatus*), is especially striking, with a cream-coloured body and striped pattern. This small animal, about 3 to 5 cm in length, appears similar at each end, except that some of the dark lines running the length of its body are continuous round the tail, and a pair of short tentacles projects from the front. The flatworm creeps with a flowing motion over rocks and stones and amongst weed and sediment, the margin of the flattened, wafer-thin body curling and rippling as it moves. Often found on wrecks or on the supports of piers, it is common in various shallow sites from Scandinavia to the Mediterranean.

## *Rock pools*

On a rocky shore, damp crevices and overhangs of ridges exposed at low tide are often studded with small, reddish, jelly-like blobs. Close by, in water-filled hollows, more of the

*Above. The green sea slug (*Elysia viridis*), seen here creeping across a mixed patch of seaweeds, feeds mainly on shallow-growing green weeds and stores their chloroplasts (microscopic structures which carry out photosynthesis in plants) intact, and still-functioning, in its body. A related species, Hedgpeth's sea hare (*E. hedgpethi*) occurs in the North American Pacific, extending south from Washington to California and Baja California.*

*Facing page. Reaching almost to the sunlit surface, fronds of thong weed (*Himanthalia elongata*) shade a pathway through a narrow gully.*

*Right. The hydroids *Aglaophenia*, here amongst branched tufts of the bryozoans *Bugula* and small, transparent sea squirts, live attached to hard surfaces or to seaweeds. On the stems, the conspicuous yellow corbulae (from the Greek for basket) are modified branchlets covered with stinging cells to protect the reproductive structures they enclose. Several species occur in Europe, and also in North America, where they are known as ostrich plume hydroids; *A. struthionides*, for example, lives along the Pacific coast from Alaska to the Gulf of California.*

same animals can be seen, but extended to their full if somewhat squat length and topped with a crown of up to 200 tentacles which retract when the animal is disturbed or out of water. This common species is the beadlet anemone (*Actinia equina*). Typically plain dark red, it can be brown or green, is occasionally speckled or streaked, and grows to about 5 cm across the base. As wave action lifts the tentacles, a ring of blue, bead-like warts is revealed around the upper rim of the column. These contain stinging cells, used by the aggressive anemone for fighting other beadlets in territorial competition, while further batteries of stinging cells in the tentacles enable it to overcome prey that even includes small fish. The beadlet anemone is widespread from the Arctic to the Mediterranean and West Africa. Its less common relative, the strawberry anemone (*Actinia fragacea*), is usually larger and has small green spots on the bright red column.

Like the beadlet anemone, the snakelocks (*Anemonia viridis*) is often seen in pools and on rocks at the water's edge. The snakelocks, though, does not usually retract its tentacles. If it is exposed above the water line at low tide, the long tentacles lie draped against the rock with all the verve of a heap of

*Above. With its broad base keeping a firm hold on the rock, the beadlet anemone (*Actinia equina*) is ready to reach out and attack any other beadlets encroaching on its patch. The necklace of stinging warts, mauve in this individual but more often blue, is the weapon with which the anemones do battle.*

*Right. A sponge-decked scorpion spider crab* (Inachus)*, living in association with the snakelocks anemone, clambers over its host's tentacles.*

*Below, and facing page, below. The tentacles of the snakelocks anemone* (Anemone viridis)*, which may grow up to about 15 cm long, sprout in sinuous abundance.*

drowned worms. In spite of what might be considered a rather unappetising appearance, the snakelocks is eaten in a variety of regional Mediterranean dishes. It generally lives in shallow, open situations, where enough light will reach the zooxanthellae – single-celled algae in the anemone's tentacles – to enable them to carry out the photosynthesis necessary for its growth. The purple-tipped tentacles of the snakelocks are green, greyish or brown, often – in the green form – handsomely decorated with a pink stripe. A slight movement may part the sinuous tentacles to reveal a small inhabitant, the scorpion spider crab (*Inachus*), its gangling legs frequently covered with orange sponges.

The tranquillity of the rock pool is often disturbed by a very common scavenger, the shore crab (*Carcinus maenas*), as it scuttles amongst stones or seaweed. When it stops, its dark olive-green or brownish carapace merges into this setting, although younger crabs may have brighter markings of whitish speckles and lines in a bilaterally symmetrical pattern. The male, at up to about 8 cm across slightly larger than the female, lies on top of her in courtship and waits for her to moult so that he can mate

*The butterfish (*Pholis gunnellus*), in close view here among the orange lobes of dead-man's fingers (*Alcyonium digitatum*), is common in a variety of shallow habitats, including rock pools. Its range extends from Europe to the western Atlantic, where it is found from Labrador to Delaware Bay.*

with her while she is still soft. The shore crab is widely distributed, occurring from Iceland and Norway to West Africa, as well as on the Atlantic coast of North America, where it is known as the green crab, and has been introduced and established in various other parts of the world, including Madagascar and Australia. It is commercially exploited on a small scale in Scotland, and its close relative, *Carcinus mediterraneus*, is fished in the Mediterranean.

A fish typical of rock pools is the slippery shanny (*Lipophrys pholis*), scaleless like other blennies and largely hidden by its dull dark colouring, though its thick lips may stand out pale as a clown's. Another visitor to rock pools is a relative of blennies, the butterfish or gunnel (*Pholis gunnellus*). This slender fish often escapes attention until an eel-like wriggle betrays its presence as the elongated body weaves amongst rocks and seaweed. It has a very small head, but its laterally flattened body, marked with a row of large, black spots along the back, may reach a length of up to 25 cm after a decade of slow growth. The butterfish owes its name to a skin slimy with mucus. Although primarily an arctic fish, it is found as far south as the English Channel and also in North America. It is not confined to shallow water but goes deeper too, particularly in winter. Females, and sometimes males, guard the eggs, which are laid in a clump, often in a crevice or a shell.

## *Protected by shells*

A variety of molluscs, and also barnacles, cover rock surfaces in the shallows. Thick shells enable them to withstand the crushing force of the waves, but those on exposed rocks also require a firm grip or permanent attachment if they are to avoid being dislodged. Mussels and barnacles fix themselves securely to rocks by means of secretions, which in mussels are fibres (the 'beard' that is removed when they are cooked) and in barnacles are a cement-like substance; limpets (*Patella*) are renowned for their ability to clamp on to rock using the powerful suction of a large fleshy foot. Chitons, the prehistoric-looking coat-of-mail shells, also have a limpet-like grip on the rock, aided by

*The candy-stripe flatworm (*Prostheceraeus vittatus*), undulating over brightly coloured encrusting sponges. Related species in North America are the horned flatworm (*P. maculosus*), which lives on the Atlantic coast at Cape Cod, and *P. bellostriatus*, which is found on the Pacific coast off California.*

their flattened shape and large surface area for attachment. Like limpets, they feed on the fine layer of algae that they scrape off rock surfaces. Acorn barnacles often crop up among patches of mussels on rocks. Unlike other crustaceans, once they are adult they live permanently cemented to hard surfaces, and catch food from the current by waving their legs through the opening at the top of the plates that cover them. Their relatives, goose (or stalked) barnacles, are rarely seen near the shore and instead drift in open seas, attached to floating objects such as buoys and boats; in a few countries they are considered a delicacy, though the long, black, slimy object you are served could fairly be described as an acquired taste.

Mussels (*Mytilus edulis*) are widespread on seashores and in estuaries, often packed densely on rocks or man-made structures such as jetty supports, or, as mussel beds, covering large stretches of sand. Like certain other bivalve molluscs, they produce fibrous threads known as a byssus to anchor themselves to a hard object. Providing one or two have found a stone or similar suitable site, others can join the colony and attach themselves to the mussels already established. A young mussel's future is partly determined by where it settles. The size and shape of the shell are affected by environmental conditions: in sheltered sites some grow to between 5 or 10 cm or more in length, while those in exposed places are stunted by comparison. Mussel shells are sometimes marked with brown rays and have a horny outer surface that is brown or bluish black. A frill of white to brownish-yellow mantle, the soft tissue of the living animal, protrudes slightly from the parted shell valves. If the mussel finds itself out of the water, the shell shuts tightly to prevent desiccation.

Dog whelks (*Nucella lapidus*) live among the mussels and barnacles on which they prey. Other intertidal and shallow-living gastropod molluscs include various top shells and cowries and the netted dog whelk (*Hinia reticulata*). Unlike their counterparts in the tropics, the cowries of north-western Europe are not especially conspicuous. Belonging to a different family from true cowries, they are so small that I have noticed them only when I have stopped to scan a patch of rock face extra carefully, with my nose inches away from it. The European cowrie (*Trivia monacha*) may grow to about 15 mm in length and has three dark spots on its shell which are missing in the slightly smaller northern cowrie (*Trivia arctica*). The oval shell is a beautiful structure, pinkish-cream with polished ridges resembling the

*Top to bottom.*
*Only the pale pattern on the shore crab* (Carcinus maenas) *stands out against a dark backdrop of seaweed. This crab can grow to about 8 cm across the carapace.*

*This starfish,* Asterias rubens, *lying on a rock encrusted with pink coralline algae, has the florid colouring typical of Norwegian members of the species; those living further south are usually paler. A similar species – the northern sea star* (A. vulgaris) *– is found on the Atlantic coast of North America, from Labrador to Cape Hatteras.*

*A rock studded with barnacles is home for two netted dog whelks* (Hinia reticulata)*; the mottled grey siphon of one protrudes from a shell worn down by the scouring action of waves and waterborne grit.*

*The European cowrie* (Trivia monacha) *moves over a ground of sponges and finds small patches of transparent sea squirts to eat. Its small siphon at one end and pin-striped foot at the back are clearly visible. Relatives of European cowries include the Californian coffee bean* (T. californiana) *and* Trivia quadripunctata.

lines of a fingerprint, but to see the shell alone is to miss much of this fascinating little animal. In life, the ridged surface is often covered by a mottled grey or brownish mantle, a thin sheet of flesh that provides camouflage but is withdrawn inside the shell for protection when the animal is disturbed. Protruding from the front is a bright orange siphon that draws water in, and if you look closely you can also see a pair of slender orange tentacles, while at the back end the foot is striped with cream lines. The European cowrie feeds on certain colonial sea squirts including the star sea squirt, and it lays its eggs inside these.

### *Seaweeds*

At low tide, long, trailing tangles of thong weed or sea thong (*Himanthalia elongata*) hang from the rocks around pools or in shallow water. As the tide rises, it buoys up the narrow, greenish-brown ribbons until they are upright. These branched fronds may grow to two metres or more, developing from a 'button' just above the holdfast that attaches the plant to the rock. Low tide exposes many seaweeds that are normally submerged, among them thick blankets of knotted wrack (*Ascophyllum nodosum*), which cover rocks and slither hazardously underfoot. The olive-green branches, which can be a metre or more long, are narrow, ribless and slightly serrated; oval air bladders at intervals along the branches help to hold them upright when they are submerged, and small side branches bear yellowish reproductive bodies. Knotted wrack grows abundantly on the middle shore of sheltered rocky coasts. It is found across northern

*Above. Calcified beads make up the branches of coral weed (*Corallina officinalis*), which are draped one over the other, obscuring the rocks on which they are growing. Outside Europe, coral weed occurs on the Atlantic coast of North America, from Long Island Sound to Newfoundland; further* Corallina *species live along Pacific coasts.*

*Overleaf. The long-bodied snake pipefish (*Entelurus aequoreus*) attempting to hide by posing in imitation of a frond of seaweed.*

*Below. Prominent tufts of bristles distinguish* Acanthochitona *from other chitons. The patchwork of colours on the shell plates of this 4 cm long mollusc are made even more vibrant by a small area of coralline algae, which is also growing on the surrounding rock along with the honeycomb structures of bryozoan colonies.*

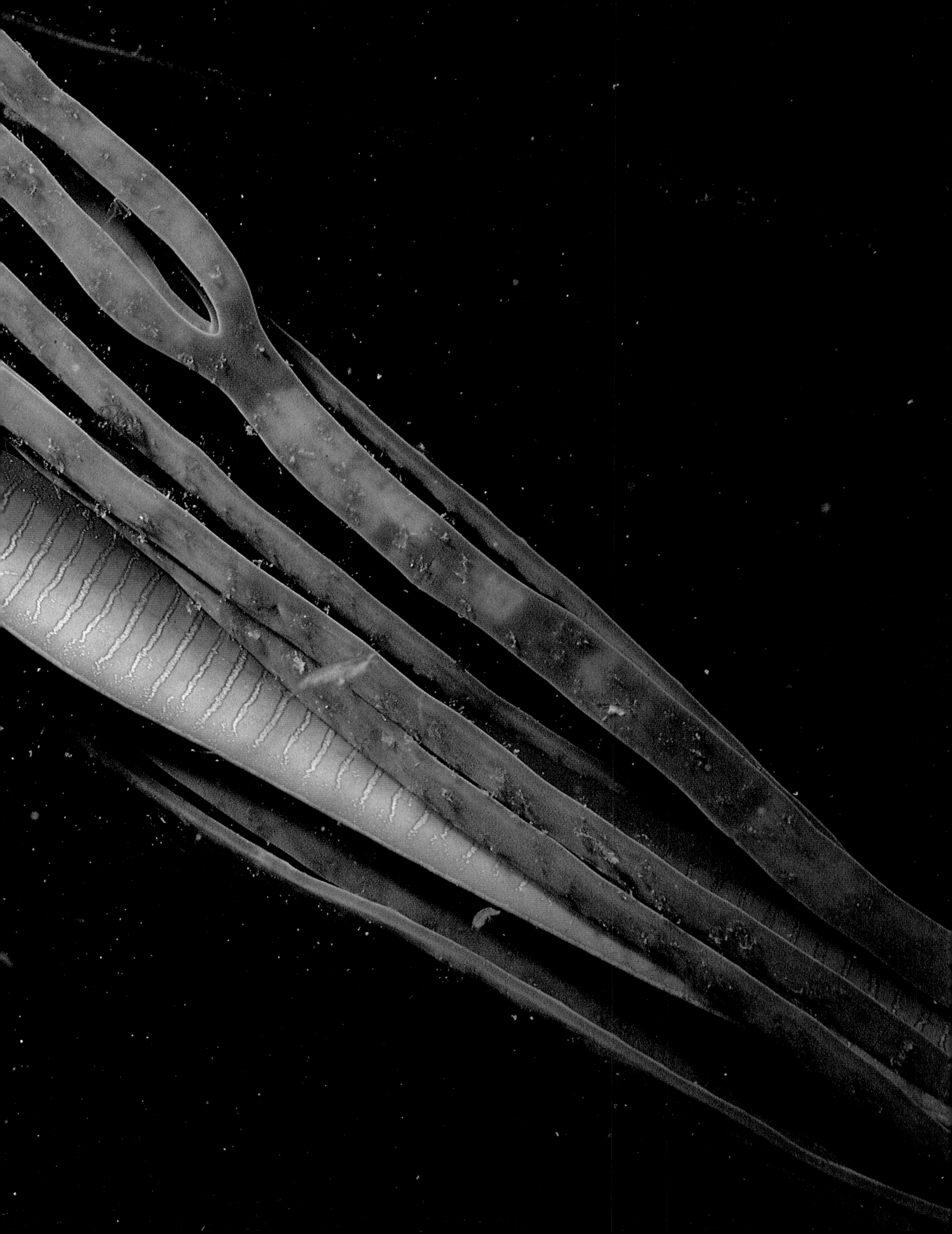

Europe and on the Atlantic coast of North America from Long Island Sound to the Arctic. Unattached plants sometimes float just offshore or drift as oceanic plankton.

Most beach walkers are familiar with bladder and serrated wracks, but certain other species of seaweed are less readily recognised. Certain kinds of red algae, for instance, have chalk skeletons that give them a hard, brittle structure. One of these, growing in purplish, fan-shaped tufts up to 10 cm long, is coral weed (*Corallina officinalis*), whose delicate branches are formed of chains of tiny, jointed segments. Coral weed covers rocks from the lower shore to the shallows, including rock pools. It, too, is a seaweed of both northern Europe and North America.

## *Hiding in the weeds*

Pipefish live at the water's edge and in the shallows generally, but they are easy to miss as they tend to align their long, narrow bodies with seaweeds or eel grass, or to wrap themselves around the fronds so closely that they are barely distinguishable from the plants. Their head shape and elongated snout, with its tiny mouth at the tip to suck in small fish fry or other planktonic morsels, are reminders that they are related to seahorses, but they are straight from head to tail, not curled. Like seahorses, they are slow swimmers with small fins, but the dorsal fin in particular allows the fish to get around as well as enabling it to hover. If it needs to move more rapidly, the pipefish twists its snake-like body from side to side to propel itself, even though

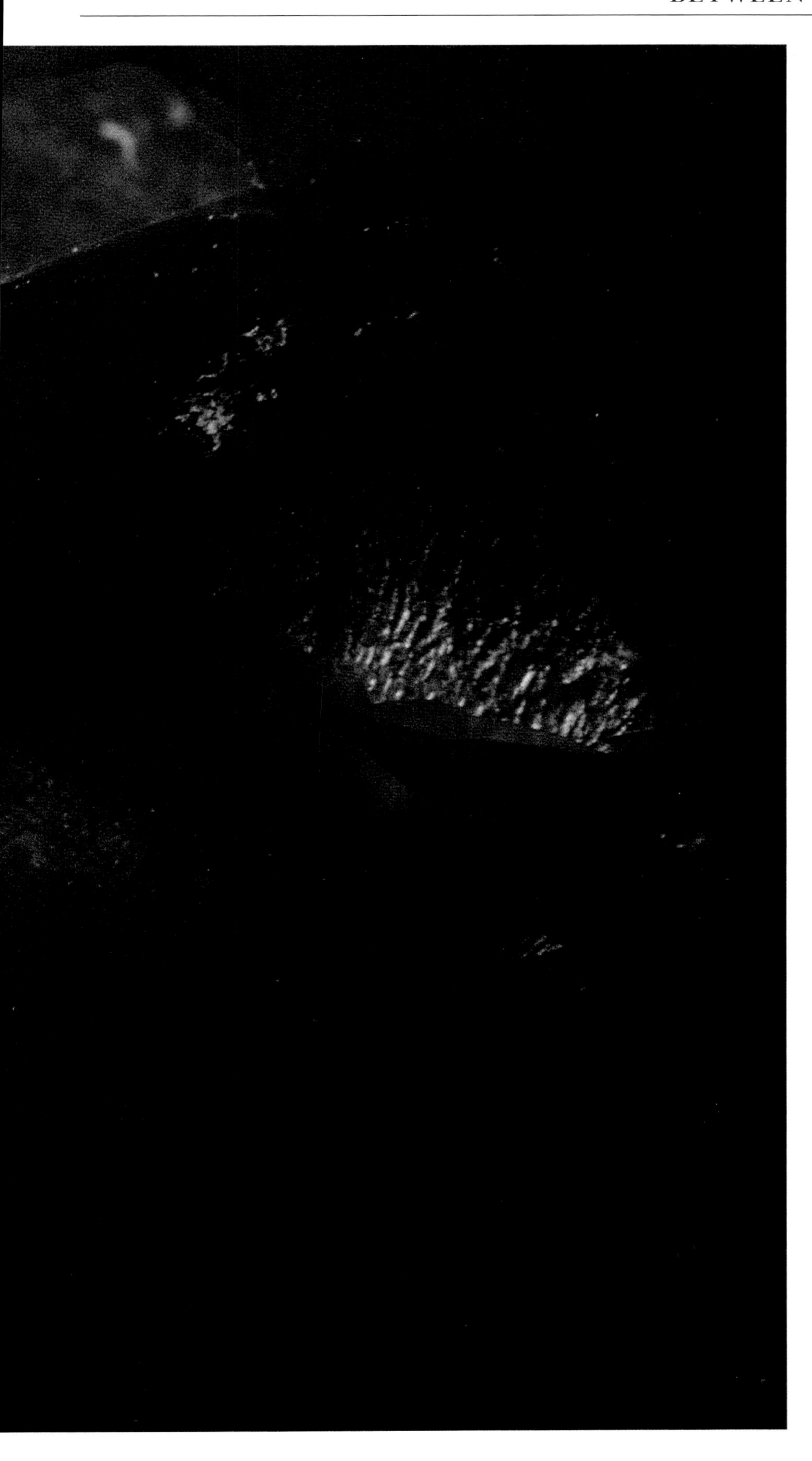

a casing of bony plates makes it less flexible than it might be. As in seahorses, the eggs are carried by the male, usually in a brood pouch, until they hatch.

Several species of pipefish can be found from Scandinavia to southern Europe, including the greater pipefish (*Syngnathus acus*), which has a particularly long snout and a distinct hump on the back of the head. Its mottled brown or olive body is prominently ringed and usually up to 30 cm long, though males can grow even larger. The snake pipefish (*Entelurus aequoreus*) is less common and, with females of up to 60 cm, the largest of the European species. Its golden or warm brown body is handsomely banded with narrow rings of pale, silvery blue, a pattern that is distinctive in the open, but can make for invisibility among weed, where it helps to break up the outline of the fish. With no pectoral fins or obvious bony ridges and only an inconspicuous tail fin, the snake pipefish has a sleek appearance. Unlike the greater pipefish, it has no brood pouch, merely a groove where the eggs lie against the male's abdomen. It often lives in deeper water than other pipefish and has also been caught near the surface way offshore, but when I have found it, whether during the day or night, it has been fairly close to the water's edge and in depths of no more than a few metres.

During May and June, the weed may well provide the backdrop for furious activity by a small member of the wrasse family. Males of the corkwing wrasse (*Symphodus melops*) pluck small tufts of seaweed and build a nest which they then guard boldly. The male, up to about 15 cm long, is always more brightly patterned than the dull brown female, but in the breeding season his colours are enhanced for courtship and rival those of tropical members of the family. The male corkwing's display is designed to attract at least one, preferably more, females to lay their eggs in his nest. The eggs stick to the seaweed, and the male then watches over his haul of potential progeny until they hatch.

Clumps of weed near the shore may also provide shelter for the common eel (*Anguilla anguilla*) at one stage in its complex life cycle. After hatching in the breeding grounds of the mid-Atlantic Sargasso Sea, the transparent larval fish drift among the plankton for about three years before reaching the coasts of Europe. There, in shallow water, they develop into yellowish-brown elvers (young eels). Most head for estuaries and swim upriver, returning to the sea several years later, when they approach maturity; others remain around the coast until they are adult and then join a migration to the Sargasso Sea.

*Above. A common eel (*Anguilla anguilla*) taking cover in a bed of green seaweed close to shore.*

*Right. A male corkwing wrasse (*Symphodus melops*) carrying nesting material – a sprig of seaweed – between his teeth.*

*Overleaf. Another male corkwing displaying the exotic colours of the breeding season. One characteristic of this variable species is the kidney-shaped blotch behind the eye.*

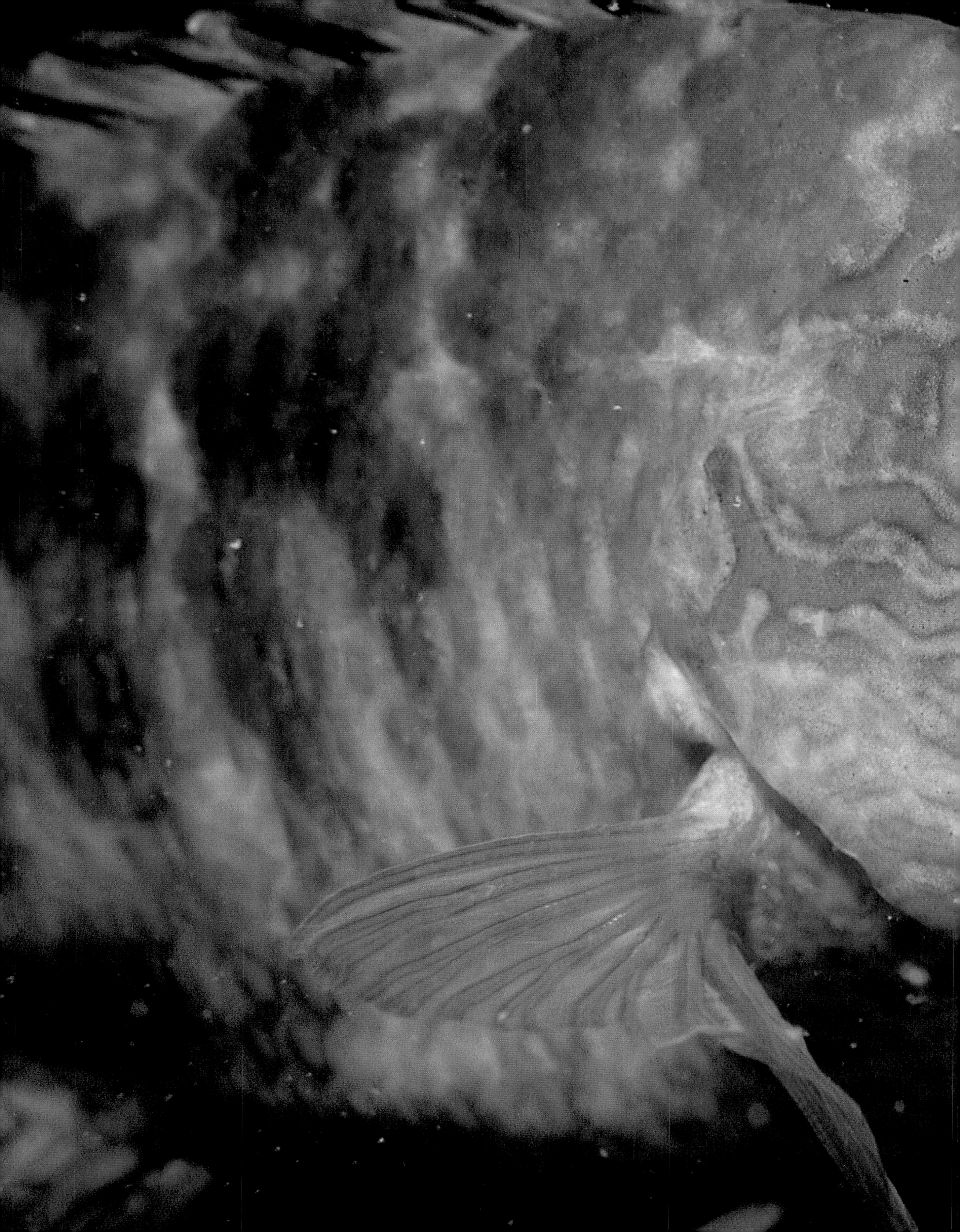

# CHAPTER TWO

# SAND AND MUD DWELLERS

On a visit to a stretch of coast on the English Channel, I dived down to the sea bed and found that somehow we had missed the expected outcrop of rocks. Instead of the interesting terrain of ledges and gullies we had hoped for, we were confronted by nothing but monotonous flatness. Sand stretched interminably into the distance, featureless except for row upon row of ripples. Reluctant to go to the trouble of surfacing to try another site, we drifted over sand, mud and shingle. Before long, our decision to explore this uniform territory was more than vindicated. The seemingly barren ground turned out to be home to an amazing variety of marine life, some visible by day and even more by night, including cuttlefish, rays and skates. Among the sand-dwellers are quite a number that are familiar from the fishmonger's display, including scallops and flatfish, though seen alive in their natural habitat they often look startlingly different.

Necklace shells (*Euspira*), the tower shell (*Turritella communis*) and the common whelk or buckie (*Buccinum undatum*) – all commonly washed up on beaches as empty shells – can be seen creeping snail-like over sand and gravel. Some bivalve molluscs, such as razor shells (*Ensis*), live buried vertically in the sand, with two short siphons just protruding above the surface to take in water bearing oxygen and food. Ragworms or nereids also burrow, many of them favouring black, muddy sand.

For animals that do not conceal themselves and lack a protective shell, an open area of sand with neither rocks nor weed can be a rather exposed environment. In many cases, camouflage provides the means of avoiding unwanted attention. Pale brown colouring and speckled patterns are commonplace among a wide range of unrelated animals: fish (from small dragonets and gobies to huge anglerfish), queen scallops, sand stars and certain other starfish, and cuttlefish. To see all of these it is best to swim slowly, scanning well ahead for signs of life. Even so, fish and other animals tend to spring up as if out of nowhere and shoot off as you draw near. To get a close view, you just have to watch where they settle and approach with care and patience.

Cuttlefish (*Sepia officinalis*), like other cephalopods, are more active and usually more approachable at night. The large eyes in the beam of my torch return my gaze with seeming intelligence. Although I cannot fathom the precise reasons for the signals the animal gives out, the almost instantaneous colour changes that pulsate over its body, making mottled or striped patterns, and the alterations in its skin texture from warty to smooth, form an astonishingly complex set of responses for a mere invertebrate. A much smaller relative, the little cuttlefish (*Sepiola atlantica*), is also common on shallow sand but less often noticed, being no more than 4 cm long.

Soon after mating and egg-laying, cuttlefish die, and their remains are washed up on the beach for strollers to find: a calcareous internal shell, the cuttlebone, polished clean by the sea, and much prized by budgerigar fanciers as a source of calcium for their birds.

## *Lying low*

Many animals' sand- or gravel-coloured camouflage is made even more effective by their flattened shape. Rays, flatfish and anglerfish project little above the sea bed they lie on, although their eyes are raised, frog-like, to allow them to keep a lookout. The ray you are most likely to see is the thornback (*Raja clavata*), which is fished in the North Sea and the Mediterranean. More often, though, divers have to content themselves with more mundane but nevertheless diverting encounters with various species of flatfish.

Flatfish lie on one side with both eyes on the other, upper side. This surreal arrangement of features contributes to a faintly ludicrous expression that varies greatly from species to species. Brill (*Scophthalmus rhombus*), a finely spotted fish with a ragged front edge to its dorsal fin, and the more commercially important turbot (*Psetta maximus*), are among those which have the eyes on the left side. Many others, including plaice (*Pleuronectes platessa*), Dover sole (*Solea solea*), lemon sole (*Microstomus kitt*), and the smaller dab (*Limanda limanda*), are right-eyed.

The plaice is one of the most familiar food fish in northern Europe. In life, however, its orange blotches make it far more

*A growth of fine seaweed, interspersed with a few barnacles, spreads over the shell (up to about 10 cm in length) of a common whelk or, as it is known in North America, a waved whelk* (Buccinum undatum)*. The animal is piebald: you can just see this from its long siphon, held aloft as it moves past patches of light-bulb sea squirts* (Clavelina lepadiformis)*.*

*In varying poses, the cuttlefish (*Sepia officinalis*) displays some of the many different colour patterns and textures its skin can assume. The rapid succession of guises reflects the changing activities of the cuttlefish as it hovers alert above the sea bed in readiness for jet-propelled action, responds to threat, or rests calmly on the sand. The zebra-like pattern is often seen in undisturbed males.*

*Overleaf. Just the faint shadow of its frilly dorsal fin against the sand, and two beady eyes, betray the presence of the brill (*Scophthalmus rhombus*), a flatfish that can grow to 60 cm or more in length. It is related to the North American windowpane (*S. aquosus*), which lives in the western Atlantic from the Gulf of St Lawrence to Florida.*

resplendent than you would guess from buying the dead fish, however freshly caught. These orange blotches distinguish it from the otherwise similar flounder (*Platichthys flesus*). Underwater, it is easy to overlook plaice because of their habit of covering themselves with a fine layer of sand. When I swim along without examining the ground ahead carefully, I often see a sudden flurry of sand as the fish takes off with undignified but prudent haste and glides away. Once disturbed, plaice tend to be wary, so to get a good view you need to spot one in time to slow down before reaching it. Providing you can avoid creating sudden water currents, the fish will remain motionless on the sea bed, seemingly unaware that its presence has been detected.

*The orange spots of the plaice (*Pleuronectes platessa*) are obvious when the fish first settles on sand. But with a few flips of the fins, and a quick wriggle, it can virtually disappear from view.*

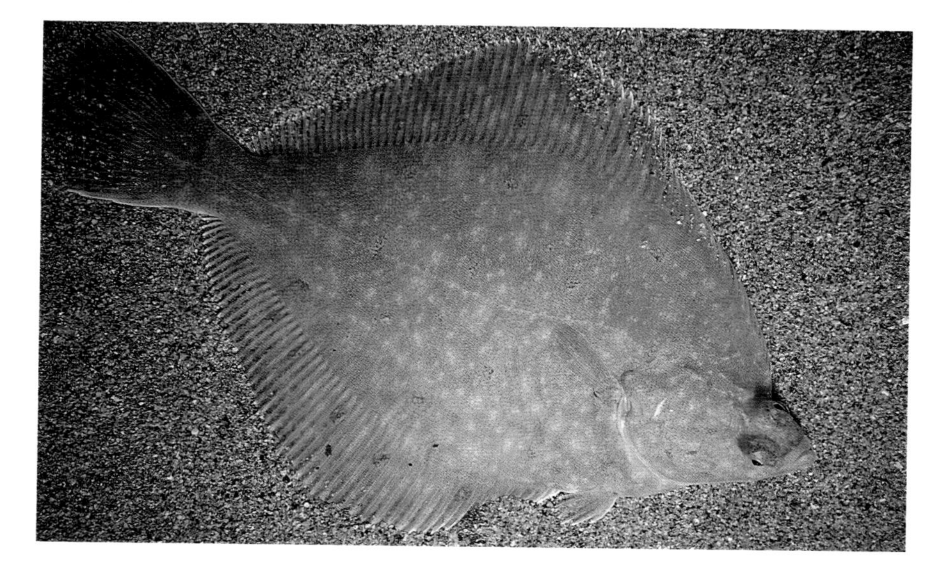

*Facing page. The lemon sole (*Microstomus kitt*) is not a true sole, but a relative of the dab, plaice and flounders. Found on various types of ground, from firm sand to rock, it is lying here on a short turf of red seaweeds and encrusting animals. Its North American counterpart is *M. pacificus*, found from the Bering Sea to Baja California and known (confusingly for the British) as the Dover sole.*

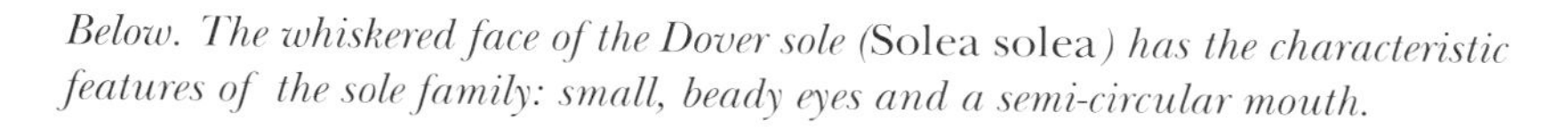

*Below. The whiskered face of the Dover sole (*Solea solea*) has the characteristic features of the sole family: small, beady eyes and a semi-circular mouth.*

One of the most bizarre fish on the sea bed is the anglerfish (*Lophius piscatorius*); like its less common relative, the black-bellied angler (*Lophius budegassa*), it will stir itself only with some reluctance when disturbed before swimming sluggishly away. A frill of fleshy flaps of imitation seaweed surrounds the mottled brownish body, forming a 'beard' across the lower jaw and blending in well with debris on the sand or gravel. The disguise works well, yet once spotted an anglerfish appears so spectacular that you wonder how you could have missed it. The enormous mouth extends right across the broad head, which accounts for nearly half of the fish's metre or so of length. Its gape is alarming, revealing many sharp teeth capable of dispatching a diet of smaller fish supplemented with crabs and lobsters or even the occasional sea bird. The anglerfish's strategy for catching these is as startling as its appearance. Above the mouth is a long, movable spine, a modified ray of the first dorsal fin, tipped with a fleshy lure which entices the unsuspecting prey close enough to be snapped up. Despite its somewhat grotesque appearance, it is fished for the market where it is sold as monkfish; although usually displayed with the head removed, it can sometimes be seen entire and gaping grimly. It is found from northern Norway and Iceland to the Mediterranean and the Black Sea.

## *Walking on sand*

Poised on the tips of independently articulated 'fingers', the gurnard moves daintily across the sand. These appendages, three on each side, are the adapted front rays of the large pectoral fins and can sense the crabs and shrimps that the fish probes for in the sediment. Several gurnard species, distinctive fish with a disproportionately large head armoured with bony plates, live in northern seas. The red gurnard (*Aspitrigla cuculus*) is one of the more resplendent, with brick-red colouring and a pale underside, and may grow to about 40 cm in length.

Another similarly sized fish, also equipped with sensory organs to detect its prey in sand or gravel, is the striped red mullet (*Mullus surmuletus*). Like other members of the mainly tropical goatfish family, it scans the shallow sea bed a small patch at a time, using a pair of long, white chin barbels, and digs out any invertebrates and small fish that it finds. The red mullet's scarlet to yellowish flanks, longitudinally striped with yellow, tend to be more blotched at night (goatfish are noted for the changes in their colour patterns according to which of their daily activities – such as sleeping or eating – they are engaged in). When the mullet raises its first dorsal fin, revealing dark bars, it may be giving a warning signal to others in the vicinity. Its northern limit is in British waters; further south, where it is more abundant, it is fished commercially and highly valued.

Many bottom-dwelling fish are easy to photograph, but dragonets (*Callionymus lyra*) can be exasperatingly perverse. Females and immature males are common on sand, mud, shingle or pebbles, often in very shallow water, but they are so well camouflaged by their dull brown blotches and sandy spots that they are likely to see the approaching diver early and move away. It is not difficult to swim within touching distance of the fish from behind or alongside, but to get a bulky camera within a few

inches of the front end of a stationary fish requires perseverance. While I focus on the fish, it watches me with protruding eyes set high on its head, and turns away each time I move in towards it, forcing me to circle until I am dizzy. After several minutes of this I am ready to forget it and look for a more co-operative subject.

On one dive, the usual, fruitless pursuit had taken place, and I had turned my attention instead to a sluggish pipefish. After a while I became aware of something else in the viewfinder; a dragonet had moved in and I was now able to focus on its head. I stayed put, and eventually two or three dragonets of various sizes became sufficiently used to my presence to go about their business as normal. While I watched, they entertained me with their fascinating feeding behaviour. Every so often one would dart forwards and make a grab at something, presumably some small crustacean or worm buried in the sediment although the action was too fast for me to be sure. The fish would then settle back, mouth slightly open, and eject a plume of mud from the small, round gill openings situated behind and almost on top of its head. Sometimes this meant that the fish was entirely enveloped in a cloud of sediment, which did nothing to improve the already poor clarity in the area. But I was so engrossed, and my air consumption at a mere three metres was so slow, that nearly three hours elapsed before I surfaced.

Although rarely encountered by divers, mature males of the common dragonet are much more striking in appearance than females. They can grow to 30 cm, over half as much again as the maximum length recorded for females, and are characterised by blue stripes and spots which are displayed during their complex courtship behaviour. When spawning begins, the pair rises towards the surface, releasing eggs and sperm. After hatching, the larval fish float as plankton, later moving down to dwell on the sea bed.

*Anglerfish (*Lophius piscatorius*) are poor swimmers, usually leaving the sea bed only to move, with an ungainly waggle, to another resting place nearby. The skin of the pale underside, revealed by the swimming fish pictured above, is soft and loose. A closely related species, the North American goosefish (*Lophius americanus*) lives in the western Atlantic from the Gulf of St Lawrence at least as far south as North Carolina.*

*Left. The anglerfish, unmistakable on a bed of dark seaweed, is more often found on coarse sand, where its colouring and 'beard' of skin flaps provide effective camouflage.*

*Overleaf. The anglerfish's long front spine, with the lure at the end of it, is laid back between its eyes, ready to be raised and twitched. Its other, spine-like rays (one seen upright here) are shorter and lack lures.*

Some of the tiniest species of fish found in northern waters are members of the goby family – many are no more than 6 cm long, and some are considerably smaller. Usually, they are an undistinguished, speckled sandy brown, blending well with their surroundings. However, when the painted goby (*Pomatoschistus pictus*) raises its dorsal fins, it reveals smart bands of black spots and, particularly in breeding males, pinkish stripes. As in other gobies and certain unrelated, bottom-dwelling fish, the pelvic fins are modified to a sucker. This is not very powerful, but helps the fish to stand its ground against wave action in shallow waters. The painted goby, common on gravel and also sometimes on harder, more stony ground, has a short life of only about two years.

## *Crabs and prawns*

The spiny spider crab (*Maja squinado*), a southerly species found from the Atlantic coast of Britain, down those of France, Spain and Portugal to the Mediterranean, is common on sandy ground and also amongst stones and seaweed. With a carapace of up to 20 cm in length, and very long hairy legs, it is the largest spider crab to live so far north and is caught commercially, but mainly eaten in continental Europe. Like other spider crabs it is quite agile, in spite of its ungainly appearance. The heavy, domed carapace is covered in short, broad-based spines and bristles, with several longer spines at the edge on each side. Its brick-red or brownish-orange colour is often obscured by encrustations of algae and various animals, including

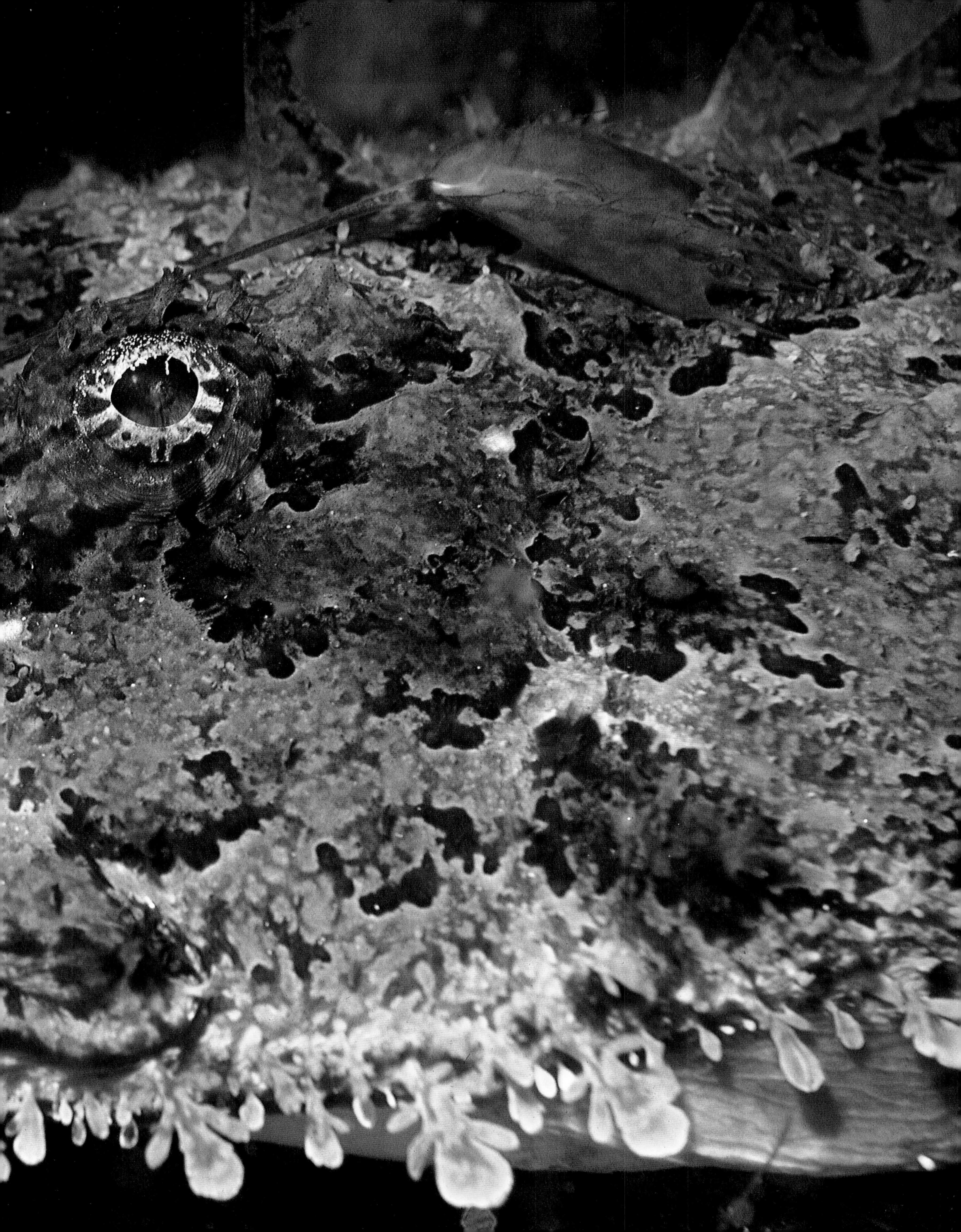

*Facing page. A spiny spider crab (*Maja squinado*) sports a fine crop of seaweed. Spider crabs decorate their carapace and legs with snippings of plants and various encrusting animals, often producing a comically dishevelled appearance, which is nevertheless an effective camouflage.*

*The blotches of the dragonet,* Callionymus lyra, *match the pebbly ground where it lives. Females grow up to 20 cm long; brightly coloured mature males may be larger. Several species of dragonet occur in Europe; elsewhere they include the lancer (*C. bairdi*), whose range extends from Florida and Bermuda to the Caribbean and the Gulf of Mexico.*

*The conspicuous white barbels of the red mullet (*Mullus surmuletus*) detect its prey of buried invertebrates. A related species, the red goatfish (*M. auratus*) occurs on the Atlantic coast of North America from Cape Cod to Florida.*

*The red gurnard (*Aspitrigla cuculus*) and other gurnards hunt in the sand like the red mullet, but instead of barbels they use finger-like rays on their pectoral fins. Many gurnards are known to make grunting noises by vibrating their swimbladder, though what purpose this has is not clear.*

*The sand goby (*Pomatoschistus minutus*) is well camouflaged on a bed of sediment and small stones.*

sponges and hydroids. In summer the crabs congregate in shallow waters and pile on top of each other in large numbers to mate. Hard-shelled males on the outside of the group protect the newly moulted, soft-shelled females at the centre.

While I was diving on the sand and gravel slope of a Norwegian fjord, a small, fat-bodied, lobster-like animal caught my attention as it scuttled across my field of view. Intrigued, I inspected it more closely and discovered that it was a hermit crab. The buff-coloured, fleshy body dotted with startling magenta blotches that I had at first supposed to be part of the crustacean belonged instead to another animal: a sea anemone. The cloak anemone (*Adamsia carciniopados*) wraps its base and column around the shell of the hermit crab (*Pagurus prideauxi*), so effectively that the shell is completely hidden from view. Although the anemone has up to 500 white tentacles, you get only the odd glimpse of these as they hang down around the crab's legs, well placed to catch stray morsels of its food. If the anemone grows too large for the shell, it can secrete a horny membrane to increase the surface area for attachment. This saves both crab and anemone from the inconvenience of moving to a new shell. The versatile anemone defends itself and its partner readily, by ejecting candy-pink threads (or acontia). These may look like festive streamers, but they are covered in nematocysts (stinging cells). The hermit's big pincer – bristly in *Pagurus prideauxi* but more coarsely granulated in the common hermit crab (*Pagurus bernhardus*) – guards the opening of the shell. Both species occur from Norway to southern Europe, and the common hermit crab also lives along the Atlantic coasts of North America.

Crustaceans seen on soft ground include the aesop prawn (*Pandalus montagui*), sometimes called the pink shrimp, which also takes up residence in wrecks and rock pools. The aesop prawn can reach a total length of about 8 to 10 cm from its tail to the tip of its long, serrated-edged rostrum or beak. Its somewhat transparent, reddish-grey colour patterned with dark red lines on the back and sides renders the animal inconspicuous in most situations, but while I was diving at night in Norway, two prawns were suddenly brilliantly illuminated in the beam of my torch: a large patch behind their heads glowed fiery red, and minute red dots sparkled in clusters over their bodies.

### *Scallops and starfish*

If you come across a shallow depression in the soft ground, it may well partly conceal the two, hand-sized valves that form the shell of the great scallop (*Pecten maximus*). The convex, right valve nestles in the hollow, and the salmon-pink to reddish-brown, flat, upper valve is usually covered with a scattering of sand and often masked by tiny encrusting plants and animals. These disguises allow the scallop some respite from a major predator, the common starfish, which clambers slowly over the shells and prizes them apart to feed on the living contents. Once detected and attacked, the scallop must resort to another defence. As soon as it senses the starfish's probing tube feet, the usually sedentary mollusc is stirred into a surprising flurry of activity. The shell valves gape wider until, with a burst of energy, the scallop propels itself off the sea bed, a short distance before sinking again. The valves clap together, the process is repeated, and the scallop swims in a series of loops.

The heavy shell of the great scallop makes its swimming motion appear more clumsy than that of the lighter and more agile queenie. In both cases this makes for a comical sight, but an effective means of escape from starfish.

Like many bivalves, including mussels and oysters, the scallop feeds by drawing water in and straining minute organisms from it. As the shell gapes slightly, you can see the tissues of the living animal, the speckled pattern of the skin-like mantle edge, ragged fringes of short and longer sensory tentacles, and numerous tiny beads that are the eyes. The sixteen or so ribs radiating across the shell valves have a rounded cross section that distinguishes this species from the Mediterranean pilgrim's scallop (*Pecten jacobaeus*).

The great scallop, found from Norway to Atlantic Spain, is fished commercially by dredging – an activity which often causes grave damage to the sea bed, denuding it of much of the other plants and animals that live there. Some scallop beds have also suffered from the adverse effects of chemicals used in nearby fish farms.

The smaller queen scallop or queenie (*Aequipecten opercularis*) is distinctly more active than the great scallop. The shell valves have more ribs, about 20, and the upper valve is more convex than the lower. The shell is often attractively patterned with speckles and zigzag bands of purplish-red, brown, orange or yellow, but as the queenie ages these markings may become obscured by encrusting animals. Many tiny blue or green eyes sparkle amongst the protruding tentacles around the rim of the shell. Although queen scallops live free on mud, sand or gravel for most of their lives, when they are very young they attach themselves to stones, shells or seaweed by tough byssus threads, similar to those produced by mussels. Queenies, found from Norway to the Mediterranean and the Canary Islands, sometimes form large colonies on the sea bed and, like great scallops, are subject to commercial fishing.

*Facing page. Unusual partners: the spotted column of the cloak anemone (*Adamsia carciniopados*) completely covers the shell of the hermit crab (*Pagurus prideauxi*). If you look closely, some of the anemone's tentacles can be seen dangling behind the crab's legs, ready to trap food particles.*

*Overleaf. A close view of the great scallop (*Pecten maximus*) reveals the living structures in detail, between the wavy edges of the shell. The numerous tentacles are sensory organs, as are the minute round eyes peeping out between them; the two, marbled bands of flesh are part of the body wall, the mantle.*

*Below. The Aesop prawn or Montagu's shrimp (*Pandalus montagui*) is one of several species of shrimp and prawn fished in northern Europe and North America; these also include its more commercially important relative,* Pandalus borealis*, often referred to in North America as the Maine shrimp, which is found on the Pacific coast of North America. Both species occur on the Atlantic coast. The Aesop prawn feeds on small invertebrates, especially the polychaete worm* Sabellaria spinulosa.

## *Sand stars*

Unlike the common starfish, most sand-living starfish show no interest in their scallop neighbours. The sand star (*Astropecten irregularis*) and the seven-armed starfish (*Luidia ciliaris*) lie on or partly buried in sand or gravel. Against this background their sandy orange colouring does not catch the eye particularly, but far more conspicuous, when not covered by sand, is the comb-like fringe of pointed white spines along the edge of each flattened and tapered arm, which are used by both starfish (together with their tube feet) to dig into the sand to conceal themselves. The seven-armed starfish is a faster mover than most of its relatives, thanks to its especially long tube feet, but neither it nor the sand star have the suckered tip so essential to the firm grip of rock-living starfish.

The sand star is normally stiff, its upper surface grainy from a covering of minute, brush-like structures known as paxillae. However, as it starts to bury itself, ripples and depressions appear over its back and arms, making it look almost as if it were deflating. The seven-armed starfish has much longer arms and is a more softly textured animal which may grow to 40 cm or more across, as opposed to the sand star's maximum of 15 cm. Both species feed mainly on other echinoderms, but while the sand star seeks out small animals such as brittle stars, the seven-armed starfish can also tackle larger prey, including common starfish. The sand star is widespread in shallow waters from Norway to Morocco, and the seven-armed starfish is found from the Faeroes to the Mediterranean and the Cape Verde Islands.

### *Embedded*

Not all animals roam freely over or through the sand; some live rooted in it. Looking uncommonly like large buttons pressed against the sand, the brownish discs of the daisy anemone (*Cereus pedunculatus*), fringed with short tentacles, are fully exposed, while their long columns are buried beneath. In contrast, cerianthid anemones' tentacles are long and fine, stretching upwards from the animal's tube of mucus and mud that is hidden in the soft sea bed. Peacock worms (*Sabella pavonina*) also live in a tube of mud and mucus, a slender construction that is usually attached to a stone, shell or other hard surface buried just below the surface of the sediment. The worm's delicately rainbow-hued whorl of tentacles – used to remove food particles from the water – can be admired until it detects sudden movement nearby and slides into the safety of its tube.

*A queen scallop (*Aequipecten opercularis*) rests on a dark piece of kelp among the debris on the sea bed, its sparkling turquoise eyes looking almost iridescent. Perhaps surprisingly, the scallop is quite capable of making a sudden leap into the water and swimming by sharply contracting the well-developed muscles that close the shell. A jet of water is abruptly expelled, and the scallop progresses with a somewhat jerky movement before opening its shell again and repeating the action. Related species – the bay scallop (*A. irradians*) and the mossy scallop (*A. muscosus*) – occur on the Atlantic coast of North America.*

*Clockwise from top left.*
*The sand star (*Astropecten irregularis*), here lying on top of the sand, can dig itself into a shallow pit for protection. A similar sand star (*A. americanus*) lives in the western Atlantic from New Jersey to Cape Hatteras.*

*The crown of tentacles (or gills) of a peacock worm (*Sabella pavonina*) fan out from a tube which, at around 20 cm (they can grow as tall as 25 cm), towers above neighbouring sea squirts (*Ciona intestinalis*).*

*Beneath the disks of daisy anemones (*Cereus pedunculatus*), their trumpet-shaped columns are buried in the sand and probably anchored to a stone about 8 cm below.*

*The seven-armed starfish (*Luidia ciliaris*) stretches out a few of its whitish tube feet to test the silty surroundings, in which the glimpses of magenta are scattered jewel anemones (*Corynactis viridis*) attached to underlying rock.*

# CHAPTER THREE

# ENCRUSTING ROCK WALLS

Rocky shorelines, unlike sandy coasts, give way to a rugged underwater terrain that may include walls and ledges, slopes or spectacular sheer cliffs. These faces have a strong scenic appeal, even if bare, and, unless they are regularly subjected to scouring by extreme surge, they are clothed in a vibrant and often fascinatingly varied covering. Plants and sessile animals (those that live attached to surfaces) jostle for space, even, in places, leaving no rock visible at all. Kelps often dominate the top of the slopes, forming distinctive communities in their own right (*see* Chapter 6). Below that zone, shaded by kelp or by overhanging cliffs, the 'turf' may be largely made up of animals, although a casual observer might mistake it for a bushy undergrowth of plants. Sedimentary rocks such as limestone are ideal for animals that need to bore holes, while the smooth, hard surfaces of igneous rocks support jewel anemones and deadman's fingers (a soft coral). Crustose coralline algae daub rocks rose-pink amongst a patchwork of encrusting life that includes sea anemones, corals, sponges, sea-squirts, hydroids and bryozoans. The majority of these animals are filter feeders, straining their diet of microorganisms from the sea's currents, and many, such as plumose anemones (*Metridium senile*), are equally at home on wrecks.

One of the confusing sights you may come across when you first dive near rocky shores are what look like pale brown tufts of seaweed sprouting from a rock, but are actually colonies of animals, notably the bryozoans *Bugula turbinata, Bugula plumosa* and other similar species, which have spirally arranged branches up to 5 cm or more long. In bryozoans, microscopic individuals in the colony are joined together in a fine, regular, mesh-like structure; the sea mat, for example, forms thin crusts on kelp.

Hydroids (sometimes called sea firs) are also plant-like in appearance, although a clump of feathers may be the first comparison that springs to mind when you see *Gymnangium montagui.* This hydroid grows to a height of about 15 cm in colonies of stiffly erect main stems with a very regular arrangement of honey-coloured branchlets. *Gymnangium* attaches itself to rocks, or sometimes shells, and occurs along Atlantic coasts, where it is one of various southern European animals that

*Left. The stiff, feather-like branches of the hydroid,* Gymnangium montagui, *stand firm against the currents and filter organic nutrients from the water flowing over them.*

*Facing page. The wall of a gully encrusted with plumose (or frilled) anemones* (Metridium senile) *and different coloured patches of sponges, over which starfish and a sea urchin* (Echinus esculentus) *can be seen crawling.*

*Left. Delicate crowns of tentacles, each about 7 mm across, in a colony of horseshoe worms (*Phoronis*). The red line just visible in the membranous bodies of some is a blood vessel. There are fewer than twenty species in this phylum, and several occur on both seaboards of North America, some with whitish, others with coloured tentacles.*

*Facing page, above. Pink coralline algae completely covers the rock on which these white sea anemones (*Actinothoe sphyrodeta*) and scattered dark tufts of red seaweed are growing. One of its relatives,* A. modesta, *sometimes known as the smooth burrowing anemone, lives on sand or gravel along the Atlantic coast of North America, from Cape Cod to Cape Hatteras.*

*Facing page, below. Trumpet anemones (*Aiptasia mutabilis*) grow to about 12 cm across their long, rarely retracted tentacles. A relative, the brown anemone (*Aiptasia pallida*) occurs in southeastern U.S.A. As in certain other anemones, when the animal is disturbed, white threads of stinging cells are discharged through the mouth.*

*The dahlia anemone (*Urticina felina*), spanning up to 20 cm, on a bed encrusted with small tube worms, and, opposite, another of the many colour variants of this species. Either this, or a very similar species, occurs in North America, but so far there has been some confusion in precise identification.*

reach their northern limit in the Irish Sea. I have seen it in some abundance around Lundy, an island in the Bristol Channel which was designated a marine reserve because of its particularly diverse and unusual fauna. Lundy is one of about 50 to 60 Marine Protected Areas (some statutory, others voluntary) that exist in the north-eastern Atlantic from Norway to the Azores.

Insubstantial as a wisp of down, a lone horseshoe worm (*Phoronis hippocrepia*) attracts little attention, but in large numbers these tiny, translucent white animals give a fluffy appearance to limestone rock. Each piece of 'down' is the roughly horseshoe-shaped fan of fine tentacles crowning a slender body that is partly hidden from view in a tube embedded in the rock. Horseshoe worms resemble miniature tube worms and have the same habit of withdrawing rapidly into their tube at the slightest hint of disturbance. They are not, in fact, true worms, but are classified as a small distinct group, the Phoronida, of which they are the sole northern European member.

## *Anemones and corals*

Rock surfaces are often liberally dotted with sea anemones such as the small white, or sometimes orange-centred, *Actinothoe sphyrodeta*, up to 5 cm across the tentacles. It is easily confused with the white variety of *Sagartia elegans* except that the latter has suckers on its column. Another species, the trumpet anemone (*Aiptasia mutabilis*), is readily recognised by the starburst of pale blue lines emblazoning its brown disc. Where it grows, it is usually fairly abundant, although it does not occur further north than south-west Britain. Sometimes seen in small groups at the base of rocks, dahlia anemones (*Urticina felina*) are as large and showy as the flower after which they are named, although they are much less conspicuous when the colourful tentacles withdraw and the column is left on view, covered with gravel and fragments of shell that adhere to the warts all over it.

Some of the most beautiful sights undersea, from northern Scotland to south-west Europe and the Mediterranean, are rock faces completely covered in jewel anemones (*Corynactis viridis*). The first glimpse as I approach such a wall is breathtaking; a vast palette of pastel hues. Each patch of colour, resulting from asexual reproduction of the anemones, is clearly delimited and can be as much as a metre across. One area may be pink, the next, pale green, cream or orange. Moving in close, I can pick out the details of the individuals, up to 25 mm across the tentacles, and their colour now seems more vivid. The tentacles have swollen ends, usually in a contrasting colour; a green specimen, for example, may have magenta-tipped tentacles while those of mainly pink or orange anemones are commonly white. The tips appear like bright beads in the beam of a torch, standing out from the more subtle luminosity of the jewel anemone's translucent body. Despite their name, these animals are in fact intermediate between corals and anemones and belong to the group known as corallimorpharians.

The name dead-man's fingers hardly evokes the beauty of the white or orange soft coral *Alcyonium digitatum*. Only when the delicate, feathery profusion of polyps retracts into the rounded branches of the colony can some resemblance to swollen spongy fingers be seen. They are widespread and sometimes abundant on exposed walls where strong currents help them to escape grazing by sea urchins.

Although most people associate corals with the tropics, and indeed reef-building corals do not occur much further north than the Red Sea, a few true corals do live in northern seas, among them the Devonshire cup coral (*Caryophyllia smithii*). Some details of its stony skeleton can be glimpsed through the

*Above. Jewel anemones* (Corynactis viridis) *thrive in shaded sites. Here, a colour variety with magenta-knobbed tentacles is growing in profusion beneath a rocky overhang. A related species, the club-tipped anemones* (C. californica) *lives in the western Pacific, mainly along the coasts from Baja California to California. The tentacles of* Corynactis *have unusually large stinging cells, grouped in spirals; these are highly effective for defence and for prey capture, but are not harmful to humans.*

translucent tissues of the living animal, particularly, when viewed from above, the circle of radiating white ridges that are partition walls (or septa). Colour variants of the cup coral include green, orange, pink and white, commonly with dark red markings flaring in a jagged ring around the mouth. The tentacles are transparent apart from minute, pigmented beads scattered over the surface, and end in a whitish knob; when extended, they give the animal a span of up to about 4 cm. Cup corals are solitary, living as scattered individuals or occasionally in small groups on rock faces, wrecks, or on stones. A pink-shelled barnacle, *Boscia anglica*, is known to live attached to the side of the cup coral, but although this association is fairly common, the barnacle is so tiny and well hidden by the coral's tentacles that I have found it on only one occasion.

## *Sponges and sea squirts*

Few people rate sponges as exciting, yet without the eye-catching splashes of colour of many of these encrusting animals, the sea bed would be a much duller place. Shades of orange and yellow predominate among the brightly coloured sponges of northern Europe. Some, including the common and widespread *Myxilla incrustans*, and *Hemimycale columella*, with its distinctive, somewhat honeycombed surface, have a growth form that is low and cushion-like. Others grow in a variety of interesting shapes, sometimes (confusingly for those trying to identify them) within the same species. If you see, for example, a cluster of orange tassels projecting from rock walls, they are likely to be a form of *Esperiopsis fucorum*, a sponge with a finely granular surface that often grows long extensions when it lives on sheltered sites, especially if there is not much surface for it to spread over. Where the current is strong, the sponge adopts a far more compact shape, making its soft, delicate structure less vulnerable to breakage.

*Right. A stony column, the skeleton of the Devonshire cup coral* (Caryophyllia smithii) *gleams white through the near-transparent outer tissues. Corals are armed with batteries of stinging cells, which can be seen here as tiny, bead-like warts all over the tentacles. The tangled undergrowth on the surrounding rock includes bryozoans. A related cup coral* (C. alaskensis) *is found in the Puget Sound area of the North American Pacific.*

One of the most striking species is the boring sponge (*Cliona celata*), which stands out from the rock face as bright yellow mounds up to a metre or more across. It lives up to its name by using acid secretions to bore into shell or limestone when young, thus gaining a secure hold on its site. This sponge is a widespread pest, causing considerable damage to oyster beds in some parts of Europe, and attacking pearl oysters in the Red Sea and the red abalone off the Pacific coast of America. Related boring sponges on tropical reefs may destroy whole colonies of coral.

Although the boring sponge is a familiar sight on my dives off rocky shores, the vivid colour always draws me to swim over and admire it. The firm structure has a surface covered in flat-topped tubercles with pores through which a current of water is drawn in. After the sponge has extracted the nutrients it needs, the water flows out from larger openings (or oscula) which have raised rims and stand in a line, sometimes – on particularly upright mounds – forming a crest. A rarer sponge, also a vivid yellow, *Polymastia boletiformis*, usually grows as rounded masses covered with large, spiky, finger-like projections and, unlike the other species, grows on rock plateaus in preference to walls.

Various sea squirts that encrust rock have a very similar appearance to sponges, most notably (as suggested by its name), *Diplosoma spongiforme*. The spectacle of extensive sheets of this

*Above. The sponge* Esperiopsis fucorum *interspersed with jewel anemones, growing on a site too current-swept for the sponge to form especially long tassels, which might grow in a calmer location.*

*Left. Two, differently coloured varieties of dead-man's fingers* (Alcyonium digitatum), *which are often seen together. Swimming in front of these soft corals are tiny, two-spotted gobies* (Gobiusculus flavescens).

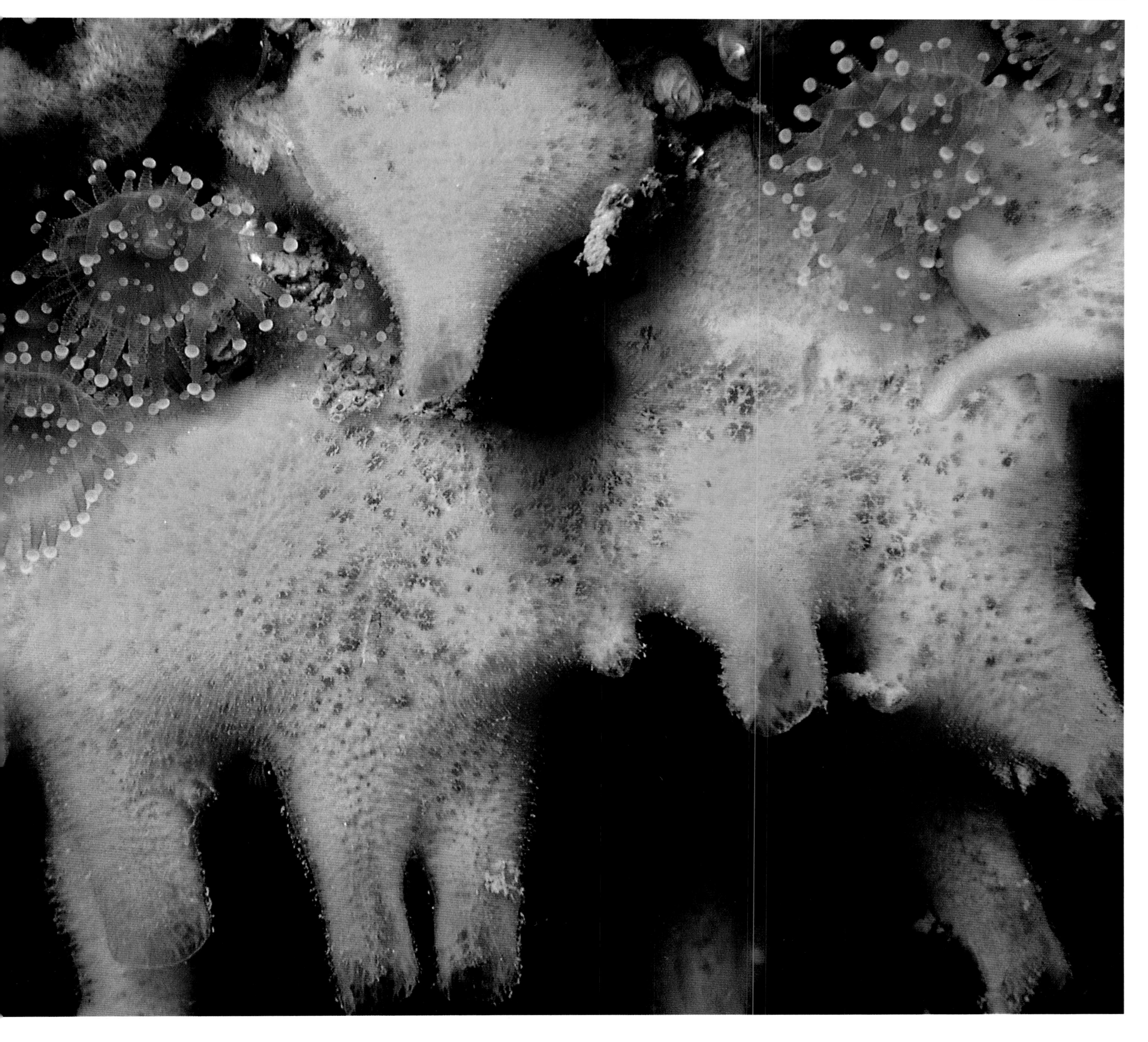

*Right. The sponge* Phorbas fictitius *has a structure of circular pits similar to that of another encrusting sponge,* Hemimycale columella. *This large patch of* P. fictitius *has spread around a small white Devonshire cup coral.*

*This page*

*Top. Prominent sponges such as the boring sponge (*Cliona celata*) sometimes offer a good vantage point for other animals to perch on. Here, a young feather star,* Antedon bifida, *clings to the sponge's water-intake tubercles. The sponge's range extends outside Europe and includes both coasts of North America.*

*Middle. The breadcrumb (or crumb of bread) sponge (*Halichondria panicea*) is common and widespread on shallow rock surfaces in Europe from the Arctic to the Mediterranean, and in North America, from the Arctic to Cape Cod on the Atlantic coast and from Alaska to Baja California on the Pacific coast. Like some other sponges, its variable form of growth can make it difficult to identify. The green colour and 'chimneys' seen here are typical of the species, and it is said to have a distinctive smell reminiscent of stranded seaweed.*

*Bottom left. The colonial sea squirt (*Diplosoma spongiforme*) covering a large area of rock, draws water in through tiny pores all over the corrugated exterior and expels it through the large holes.*

*Bottom right. Club-shaped colonies of the sea squirt* Aplidium punctum, *about 3 cm high, may each contain 40 or more individuals. They live attached to rock or, as here, seaweed.*

*Facing page.*

*Top left. This small colony of the sea squirt* Stolonica socialis *is attached to the rock wall of a gully. Each individual has two siphons at the top and stands about 2 cm tall.*

*Top right. A cluster of purse sponges (*Scypha ciliata*), each about 3 cm high. The circlet of needle-like spicules surrounding the top opening (or osculum) is part of the sponge's calcareous skeleton. Similar* Scypha *species are found on both coasts of North America.*

*Below. Beneath the kelp zone at the top of the rock slope, orange and white varieties of plumose anemones (*Metridium senile*) grow among clumps of dead-man's fingers. Crawling between them are the starfish* Asterias rubens.

species, forming a mottled grey or pale brown and whitish layer over rock surfaces makes it hard to envisage sea squirts as anything other than simple life forms, but they are in fact complex animals related to vertebrates. Many of them are colonial species consisting of numerous individuals beneath a shared outer layer (or test), which in *Diplosoma spongiforme* is furrowed, as well as being peppered with small, white-rimmed pores for water intake and fewer, but larger, outlet holes.

Other encrusting colonial sea squirts include white or pink patches of *Aplidium proliferum* and the soft, pinkish orange clusters of *Morchellium argus*. A characteristic feature of the latter – four minute red dots at the top of each translucent vase-shaped individual – can be seen through a close-up lens. Several species of sea squirt look like orange or red grapes and are often packed densely under rock overhangs. The most common of these are the gooseberry sea squirt (*Dendrodoa grossularia*), and two others, *Stolonica socialis* and *Distomus variolosus*. Some of these sea squirts, like many other animals that encrust rock, also live on wrecks, and further species are covered in Chapter 5.

# CHAPTER FOUR

# ON THE OPEN REEF

*Facing page.*
*Top. A strongly pigmented young pollack* (Pollachius pollachius) *sheltering near the wall of a shallow gully. On the rock behind the fish are the white lobes of dead-man's fingers* (Alcyonium digitatum) *and jewel anemones* (Corynactis viridis). *The closely related saithe or coley* (P. virens) *is sometimes also known as pollack in Atlantic North America, where it occurs in Greenland and from the Gulf of St Lawrence to New Jersey.*

*Bottom. Spindly arms wriggle between sturdier appendages as the common sunstar* (Crossaster papposus) *encounters a bed of brittle stars* (Ophiothrix fragilis).

The plants and animals that encrust rock provide plenty of shelter and an abundance of food for a wide range of animals that roam over rocky reefs. Some rock dwellers, such as sculpins and other bottom-living fish, are largely sedentary and content to lie in wait for their prey. Others, including starfish and sea slugs, crawl or creep over the rock face, and active fish such as cuckoo wrasses and ballan wrasses swim close to it. Some members of the cod family – pollack, bib, and the smaller poor cod (*Trisopterus minutus*) – shoal around rocks, as they do around wrecks (*see* Chapter 5).

Living in the open can be risky, but most animals have defensive strategies to bring into play against predators. Like the sand dwellers, many are camouflaged by a very similar colour to that of their surroundings. This is true of some sea slugs, although others go to the opposite extreme with flamboyant warning coloration which signals their distasteful or poisonous nature. A covering of large or small spines protects most echinoderms, including starfish, brittle stars and sea urchins, and also some slow-moving fish such as sculpins and lumpsuckers. Sea urchins are rarely seen on exposed walls and live instead on more sheltered ledges or in between rocks, mainly grazing on low-growing algae, but also feeding on animals such as hydroids

*This page, clockwise from top right.*
*A blood star or bloody henry* (Henricia oculata) *clambers over a colony of dead-man's fingers, which react to the contact by retracting their feathery polyps. A close relative, also referred to as a blood star,* Henricia sanguinolenta, *occurs in North American waters.*

*A rigid cushion star* (Hippasteria phrygiana) *crawling over rock covered with the curled tubes of small worms. The small cream disc on the animal's upper surface is its madreporite (the sieve plate which controls water intake in starfish and other echinoderms). It also occurs on the Atlantic coast of North America, where it is known as the horse star, and related species live in the northern Pacific.*

*This purple sunstar* (Solaster endeca) *on a silt-covered rock is a particularly flamboyant individual; most of those I have seen are somewhat paler. Sometimes known in North America as the smooth sunstar, it is 20 cm in radius and preys chiefly on small starfish (sea stars) and sea cucumbers.*

*Above. Competing for resources on a crowded bed, some brittle stars (*Ophiothrix fragilis*) find extra space on the lobes of dead-man's fingers, using the spines of one or more arms to anchor themselves, and extending others for filter feeding.* Ophiothrix *species occur in many parts of the world; others in northern seas include the daisy brittle star (*O. aculeata*), common on Atlantic coasts of North America from the Arctic to Cape Cod.*

and sea mats growing on rock surfaces. Any bare patches they leave behind them are rapidly recolonised.

Crabs and other crustaceans are common around rocks, but they tend to venture into the open mainly by night, seeking the safety of crevices to rest in by day (*see* Chapter 7). Hermit crabs generally live on softer ground, but certain species can be seen scuttling over rocks.

## *Stars of the reef*

Starfish such as the blood star or bloody henry provide a striking splash of colour to many exposed rock walls. The names apply to two species that are almost impossible to tell apart, *Henricea oculata*, which is found from the Bay of Biscay to its limit along the west coast of Britain, and its more northerly counterpart, *Henricea sanguinolenta*, from northern Britain to Scandinavia and across the Arctic Circle to Greenland and the North Pacific. The top surface of both starfish is usually purple or deep rose, but they can be scarlet, orange, yellow or brownish and are sometimes blotched instead of plain. They have stiff, rounded arms, rough-textured above, with scattered groups of small blunt spines, and grow up to 12 cm or more across.

Another starfish found on rock, as well as on gravel beds, is the rigid cushion star (*Hippasteria phrygiana*), a species rare as far south as Britain and more common in Norway and Iceland. It is

*Right. A black brittle star (*Ophiocomina nigra*) crawling over a starfish,* Asterias rubens. *This crowded bit of territory also accommodates a turf of small bryozoans,* Bugula, *and patches of encrusting sea mats.*

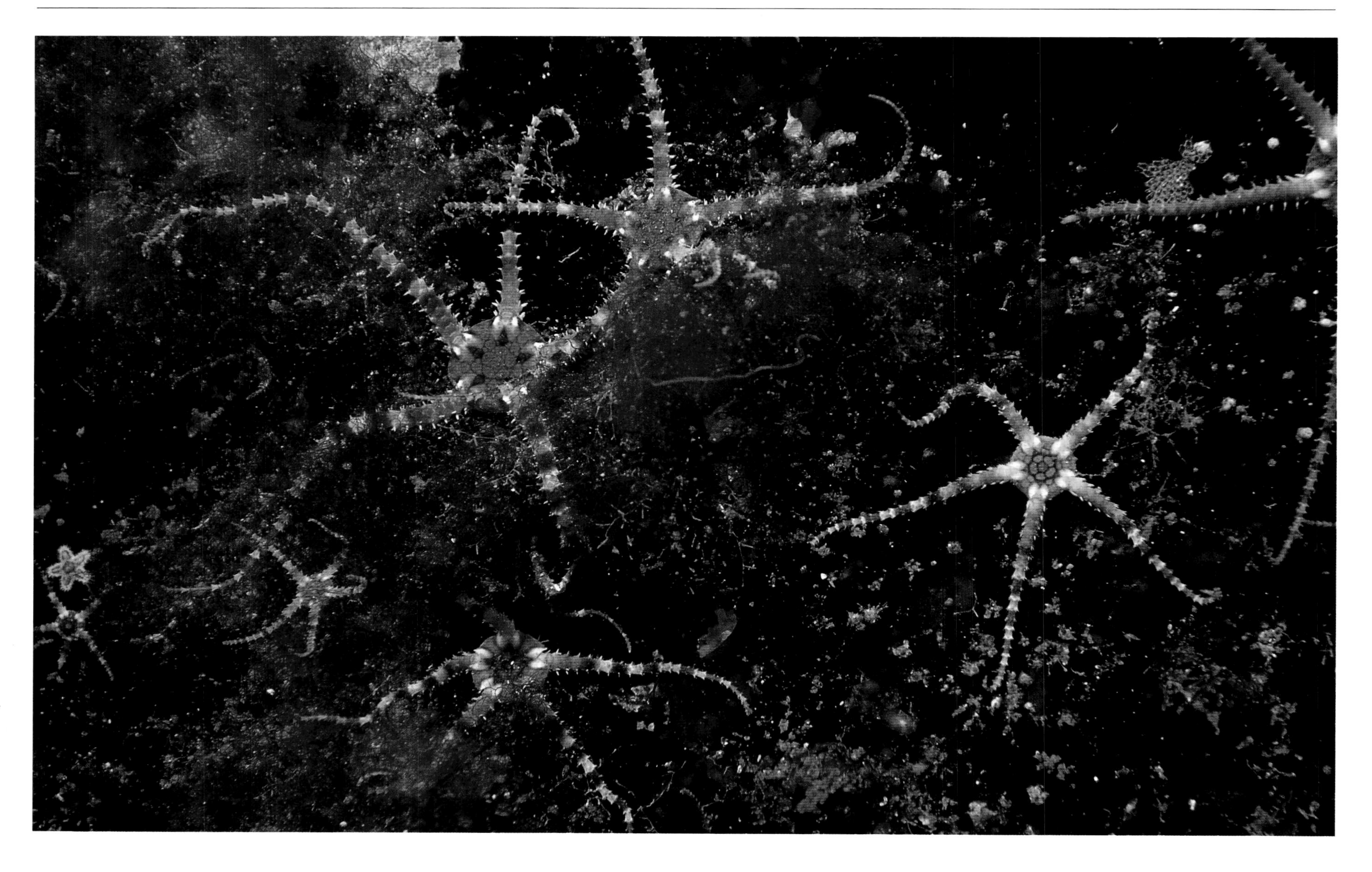

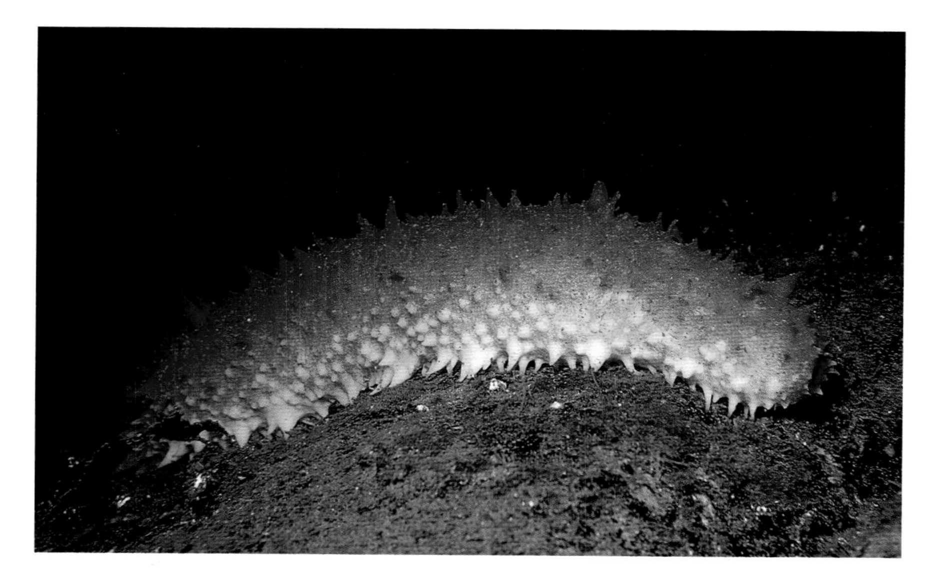

*This page, top to bottom.*
*Various sizes of* Ophiura *brittle stars draping their narrow arms over red seaweed on a rock. These starfish are also sometimes found on sediment. European species include* O. affinis, *and* O. robusta, *which also occurs in the western Atlantic from the Arctic to Cape Cod; other North American species include the circumpolar* O. sarsi, *found on shell sand on Atlantic and Pacific coasts.*

*A cotton spinner (*Holothuria forskali*), unusually yellow for this species of sea cucumber, reveals some of the small tube feet on its underside as it walks over a well-encrusted boulder. On the right is a colony of the sea squirt,* Stolonica socialis.

*Like many other members of its family, the Stichopodidae, this large sea cucumber (more than 30 cm long) has spiky papillae (projections which are modified tube feet) on its back.*

*Facing page*
*Perched on the steep wall of a fjord, a feather star,* Antedon petasus, *spreads its banded arms into the current. At the base of the animal, small, finger-like cirri grip a prominent nodule of rock. An unidentified deep-water species of* Antedon *has been recorded in North America, off the Pacific coast.*

larger, up to 20 cm across, and armoured with a hard exterior which has white, knob-shaped spines on the upper surface. A row of plates round its edge makes it look a bit like a pie crust.

Most starfish have five arms but sunstars have more (between about eight and thirteen) radiating from their large, dinner-plate-sized bodies. Two species, the common sunstar (*Crossaster papposus*) and the purple sunstar (*Solaster endeca*), are widely distributed from northern Europe to the Atlantic and Pacific coasts of North America, though the plump-bodied purple sunstar tends to live in more northerly waters and is also marginally the larger of the two, with a diameter of 30 to 40 cm. The tips of its flexible arms sometimes curl up to reveal a glimpse of the pale underside, and a dense covering of very small

*Above. In the nudibranch sea slug* Janolus cristatus, *the digestive gland shows as dark lines running up the centres of the lobes all over its back (the cerata). At its front, the pair of rhinophores with a ribbed appearance are sensory tentacles. Several other* Janolus *species are known, including a few on the Pacific coasts of North America.*

*Facing page.*
*Top. This large ribbon of nudibranch spawn, laid amongst jewel anemones,* Corynactis viridis, *may contain 50,000 eggs or more. The form taken by the spawn varies from species to species; in some, for example, it is a thin, coiled string, in others, a small, kidney-shaped bag.*

*Bottom left. In the sea slug* Okenia elegans, *the sensory rhinophores are large and boldly coloured, probably indicating their unpalatability to potential predators. Other species occur in Europe, and North American relatives include the rainbow slug (*O. sapelona*) on the Atlantic coast.*

*Bottom right. The head-to-tail mating position of a pair of sea slugs,* Polycera faroensis, *brings their reproductive organs – on the right side of each animal – into contact. Outside Europe, several related species occur in North America, including the banded polycera (*P. zosterae*), from British Columbia to California, and* P. dubia, *from Labrador to Long Island Sound.*

*Left. This leafy dendronotid or bushy-backed sea slug,* Dendronotus frondosus, *creeping over the crust of coralline algae on top of a boulder, did not have to travel far to find food; on most of the vertical faces of rocks at this site there were large patches of the hydroids it eats: oaten pipes. This sea slug is widespread in northern Europe and in North America is found from the Arctic to New Jersey on the eastern seaboard, and from Alaska to California in the west.*

spines on the upper side gives it a smoother appearance than the common sunstar, which sprouts clusters of bristle-like spines.

Both species feed on smaller echinoderms and also on molluscs. The common sunstar is often found on brittle-star beds where it is probably preying on these smaller relatives of starfish. In some places, usually where the surface is stony, brittle stars carpet the ground in vast groups, their five, spine-covered arms writhing over each other or raised into the current to catch small plankton and other organic particles. On rock faces, brittle stars crawl actively by flexing their long, thin arms, and are often seen wrapped around the sedentary animals, such as dead-man's fingers, which live there. In a desperate bid to escape predators, brittle stars will, as a last resort, shed their arms from the small central disc. Fortunately appendages can be regenerated in a matter of a season, though the starfish may then look a little misshapen. Unlike other echinoderms, brittle stars use their tube feet not for walking, but for passing food to the mouth. One of the larger species, the common brittle star (*Ophiothrix fragilis*) has arms up to 10 cm long, typically with alternating pale and darker bands of rose red, orange or grey. Five radiating rows of small spines adorn the upper surface of its pentagonal disc, differentiating it from the sturdier black brittle star (*Ophiocomina nigra*), which has a smooth disc. Both species occur from Norway to the Azores and the Mediterranean, but the common brittle star is also found much further south.

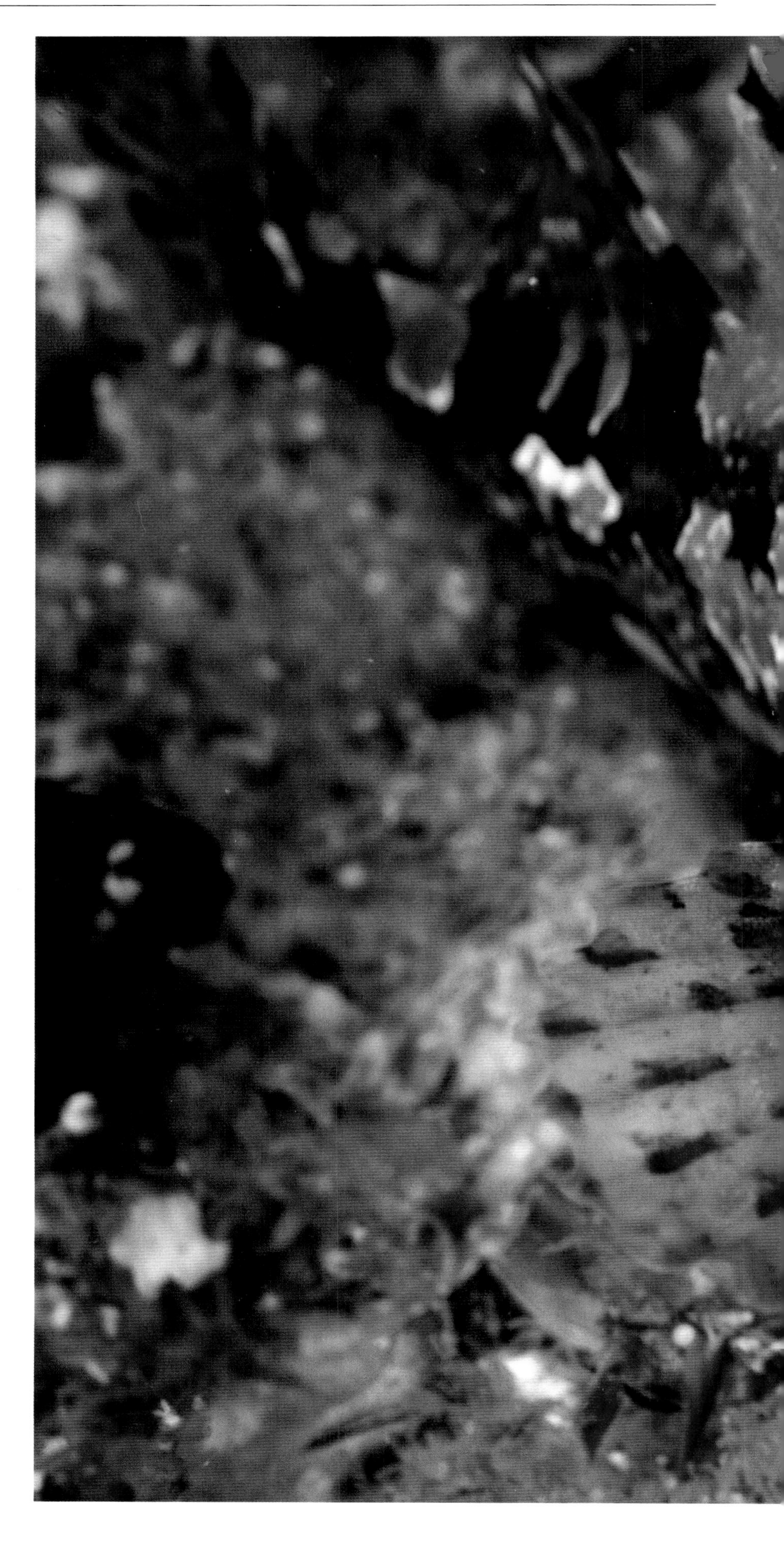

*Left. A long-spined sea scorpion* (Taurulus bubalis), *barely 5 cm long, pausing among jewel anemones which have retracted their tentacles.*

*Above. A larger individual, about 10 cm long, shows the white mouth barbel that characterises* T. bubalis. *Its colouring is exceptionally vivid, yet sufficiently mottled to enable the fish to be overlooked in the jumble of surrounding hues.*

Certain smaller brittle stars (*Ophiura*) live mainly on rock, but others are more common on mud or sand.

Unlike starfish and brittle stars, the feather stars – another group of echinoderms – generally anchor themselves to the rock by their basal appendages (or cirri). They are able to swim, but rarely choose to launch themselves into the perils of mid-water, preferring merely to let their feather-like arms reach out from rock walls and trap organic particles from the current. The northern species *Antedon petasus* from Scotland and further north is more handsomely banded than its smaller relative, *Antedon bifida*.

## *Cotton spinner*

Another group of echinoderms, the sea cucumbers, mostly remain immobile and hidden in crevices or buried, except for their crown of branching tentacles, but one of the largest and most common in northern seas lives out in the open. The cotton spinner (*Holothuria forskali*), up to about 25 cm in length, resembles a giant caterpillar. Its soft body, black or mottled with yellow, has a leathery surface with pointed protuberances, sometimes tipped with white, all over the back. Numerous tube feet on the underside enable the sea cucumber to crawl slowly over the rock face, while small tentacles sweep a mixture of sediment and organic matter into its mouth. The name 'cotton spinner' refers to the animal's defensive habit of ejecting a tangle of sticky white threads from glands near the anus. It lives on exposed sites along the Atlantic coasts of Europe and also in the Mediterranean.

## *Sea slugs*

Among the many molluscs living on rocky sites, nudibranch sea slugs are some of the most colourful, often impressively brightly patterned in contrast to their surroundings. Even so, they are easily overlooked amongst the profusion of fronds, branches and debris covering the reef. Think small to find them! Sea slugs have diverse feeding habits but many prey on particular animals that encrust rock, including sponges, sea anemones, hydroids, and soft corals such as dead-man's fingers, so a little research into their behaviour can help in tracking them down.

Despite their beauty, few sea slugs are known well enough to have earned common names, so that one of the most magnificent of more than 100 northern European species goes solely by the name of *Okenia elegans*. It is rarely seen, despite its conspicuous colouring and relatively large size of up to 8 cm, and when I have come across this spectacular animal it has been the highlight of my dive. The two occasions when I have been lucky enough to find one were both in West Wales, an area rich in sea slug species, but it is also known off south-west England, Northern Ireland, Denmark, in the Bay of Biscay and at a few sites in the Mediterranean. A fringe of orange or yellow, finger-like projections extends from the front of the head along the peach-pink back, and a matching trim surrounds the large muscular foot with which the animal creeps over the sea bed. A pair of red-tipped rhinophores (sensory tentacles) are held erect on the head, and the plume of gills situated further back is orange or rose and yellow. The sea slug burrows into and eats sea

*Right. Nestling in this colourful setting of jewel anemones, a sea snail (*Liparis liparis*) clinging to a rock face by its ventral sucker almost escaped my notice. The fish feeds mainly on shrimps and other small crustaceans. North American relatives include the showy snailfish (*L. pulchellus*) in the Pacific from the former USSR to Alaska and California, and the Atlantic snailfish (*L. atlanticus*) from Labrador to Connecticut.*

*Facing page. The fatherlasher (*Myoxocephalus scorpius*) is more heavily built than the long-spined sea scorpion and capable of tackling larger prey, such as the spider crab* Hyas.

squirts (among others, *Ciona* and *Molgula*) that live attached to the rock face and so have no means of escape.

While diving during the summer, I have sometimes seen pairs of white sea slugs side by side and head to tail. This intriguing behaviour is the usual way of mating for *Polycera* and other sea slugs, because they are simultaneous hermaphrodites (male and female at the same time), and each of the pair passes sperm and receives it from the other. The eggs of *Polycera* are laid between May and September, and the white, curled bands, each containing perhaps 10,000 eggs, are attached to the bryozoans (sea mats) on which the sea slugs feed. *Polycera faeroensis* lives mainly around the western coasts of Sweden, the Faeroes and the British Isles, feeding on the bushy spiral tufts of *Bugula* and

*Below. A Norwegian topknot (*Phrynorhombus norvegicus*) lying on a rock ledge holds its transparent dorsal fin erect above a colourful patchwork of large scales.*

certain other bryozoans. It can grow to 45 mm in length, and has yellow rhinophores, gills, and projections around the front of the head similar to those of its relative *Polycera quadrilineata*, a species found on kelp.

Also likely to be found feeding on the bryozoan *Bugula* is *Janolus cristatus*, a delicately beautiful animal that may grow to 75 mm in length. Most of its body is hidden beneath a mass of swollen, almost transparent lobes or cerata, each tipped with an iridescent sheen of bluish-white. This sea slug is found from Norway to the Mediterranean; I have sometimes seen it in impressively large numbers, particularly during June in West Wales.

Various hydroids are worth inspecting for sea slugs such as *Dendronotus frondosus*, a large, narrow-bodied animal up to 10 cm long. As it grows, it changes its diet from one group of hydroids, including squirrel's tail (*Sertularia argentea*) and sea cypress (*Sertularia cupressina*), to another, favouring as an adult oaten pipes (*Tubularia indivisa*). Its colour also changes with age. Very small individuals are whitish; larger ones are vividly speckled

and streaked with yellow, orange, red and brown, while those longer than about 3 cm have a more uniform, duller colouring. Long, branched projections around the head and in pairs down the back become increasingly elaborate with age, and give this sea slug a distinctive appearance. Other hydroid feeders include many sea slugs such as *Coryphella browni* and the tiny magenta *Flabellina pedata* – members of a group called Aeolidacea that have the remarkable ability to store stinging cells from the hydroids that they eat and use them for their own defence.

Spawn masses of sea slugs are sometimes seen near the adults that have laid them. They vary from species to species, some like a twisted thread of beads, others in the form of a coiled, lacy ribbon.

### *Scorpions and sculpins*

Bony plates and spines on the head give scorpionfish and sculpins a rugged appearance. Many of the scorpionfish found in warm waters are armed with venom-bearing spines and are highly dangerous, but those in northern seas are harmless. The Norway haddock or small redfish (*Sebastes viviparus*), which hovers in scattered groups around rocks and seaweed, is an edible scorpionfish, though not marketed as extensively as the redfish (*Sebastes marinus*) to which it is related.

Sea scorpions (which are sculpins – a separate family from scorpionfish) are scaleless, bottom-living fish. Their mottled coloration is variable, commonly brown or greenish but occasionally a magnificent purplish red, tending to match the predominant hues of the rocks and weed where the fish lie. With such excellent camouflage they can afford to be sluggish, awaiting their unsuspecting prey of small fish, crabs and shrimps. A small white barbel at each side of the mouth distinguishes the long-spined sea scorpion (*Taurulus bubalis*) from the thickset fatherlasher (*Myoxocephalus scorpius*), known also as the bull rout or short-spined sea scorpion. Female fatherlashers may grow to a length of 30 cm, which is almost twice the maximum size of the long-spined sea scorpion. Both species are fairly widespread in northern Europe, and a subspecies of the fatherlasher occurs off Greenland and the Atlantic coast of North America. The long-spined sea scorpion lives in shallow

*Above. A female cuckoo wrasse (*Labrus mixtus*) swimming over a boring sponge (*Cliona celata*) and a spiny starfish (*Marthasterias glacialis*) on boulders at a depth of about 25 metres.*

*Right. The splendid breeding colours of the male cuckoo wrasse.*

*Facing page. The John Dory (*Zeus faber*) displays its highly distinctive features: small, sharp spines along its edge, a dorsal fin with long spines, and a bull's-eye spot on its flank.*

waters where it is common in kelp gullies as well as on more open sites.

### *Other reef fish*

Flatfish skimming over ledges rather than sand are an unexpected sight, but a few, the topknots (whose eyes are on their left side, unlike the plaice and the sole which have them on the right), live on rocky ground and feed on the small crustaceans and fish that they find there. The topknot (*Zeugopterus punctatus*) grips surfaces by suction and is able to cling underneath or on the side of rocks. Another species, the Norwegian topknot (*Phrynorhombus norvegicus*), a very small flatfish no more than 12 cm long, has large scales on the eyed side and is often colourfully patterned, rather like an animated carpet.

A variety of other fish live in rocky areas off Atlantic coasts. On one occasion, while diving in Scotland off St Kilda, I noticed a strange little fish pressed against the rock face. It resembled a tadpole in shape and in its slimy scaleless skin, but the orange colouring almost matched that of the jewel anemones, among which it was nestling. This fish was the sea snail (*Liparis liparis*), one of several species that use a sucker on their underside (modified from the pelvic fins) to attach themselves to smooth rock. Sea snails can grow to 12 cm or more but are commonly only about half that length. A close relative, Montagu's sea snail (*Liparis montagui*), tends to be found in shallower water, often in pools on the shore.

Rounding the rocks below the kelp zone, I often find myself facing the long snout of a cuckoo wrasse (*Labrus mixtus*). Males are particularly inquisitive about divers and may swim repeatedly almost up to your mask and circle round behind you

*Below. The Norway haddock (*Sebastes viviparus*), found from Iceland to Ireland, is typically about 25 cm long. Its relative, the redfish (*Sebastes marinus*), is usually more than twice that size and is an important commercial fish in both northern Europe and Atlantic North America.*

before reappearing. Sometimes I have noticed a wrasse hovering above me as if intrigued by the fine stream of bubbles I am producing. Such behaviour may be a product of curiosity but, during the breeding season, males are fearless in defence of their territory and nest site and can nip at fingers quite aggressively. At this time, mature males display bright hues of blue and gold or orange. During courtship and spawning, the male has been observed to develop very rapidly an additional sexual signal, a white patch on the forehead. Cuckoo wrasses, found from southern Scandinavia to the Mediterranean and North Africa, are long-lived, growing slowly to as much as 35 cm in length. As in many other wrasses, nearly all the fish start out as females and some eventually change sex to become breeding males. Females and juveniles lack almost all the blue markings and are rosy orange with three dark spots alternating with white along the back towards the tail.

A rather grotesque fish, the John Dory (*Zeus faber*), may be seen around rocks, particularly where these border sand, and also around wrecks and amongst seaweed. It grows to about 40 cm, or occasionally more, and appears to be largely head and mouth, with an intimidating crest of dorsal, spine-like fin rays behind. Legend has it that the distinctive dark spot on each side of the laterally flattened body is the fingerprint of St Peter, who held the fish to take a coin from its mouth. This slow swimmer is usually seen hovering a few inches above the sea bed. It hunts smaller fish, often by night, relying on stealth to get close enough to extend its large, hinged jaws and grab the prey. The dappled, dull yellowish markings on the John Dory's flanks blend in well with patches of sand and weed amongst the rocks, and in head-on view, as it stalks towards its prey, the narrow body seems almost to disappear. Despite being lean and spiny, the John Dory is prized for its flavour and is fished by long line, and sometimes also by trawling. Although it is primarily a Mediterranean species and is uncommon in colder waters, it does sometimes spawn off Europe's Atlantic coasts and has been recorded as far north as southern Norway.

*Above. The common hermit crab (*Pagurus bernardus*) on a ground of crustose coralline algae, is found both in Europe and on North American Atlantic coasts.*

*Facing page. Pausing on a gaudy background of pink coralline algae, this spider crab (*Hyas*) is wearing a coat of bryozoans and small invertebrates so luxuriant that there is very little of its carapace visible enough to confirm its identity. It probably belongs to the species* Hyas coarctatus, *which, in common with the related sea toad or toad crab (*H. araneus*) is found on the Atlantic coasts of northern Europe and America. Its living covering is attached by means of Velcro-like hooks on its carapace.*

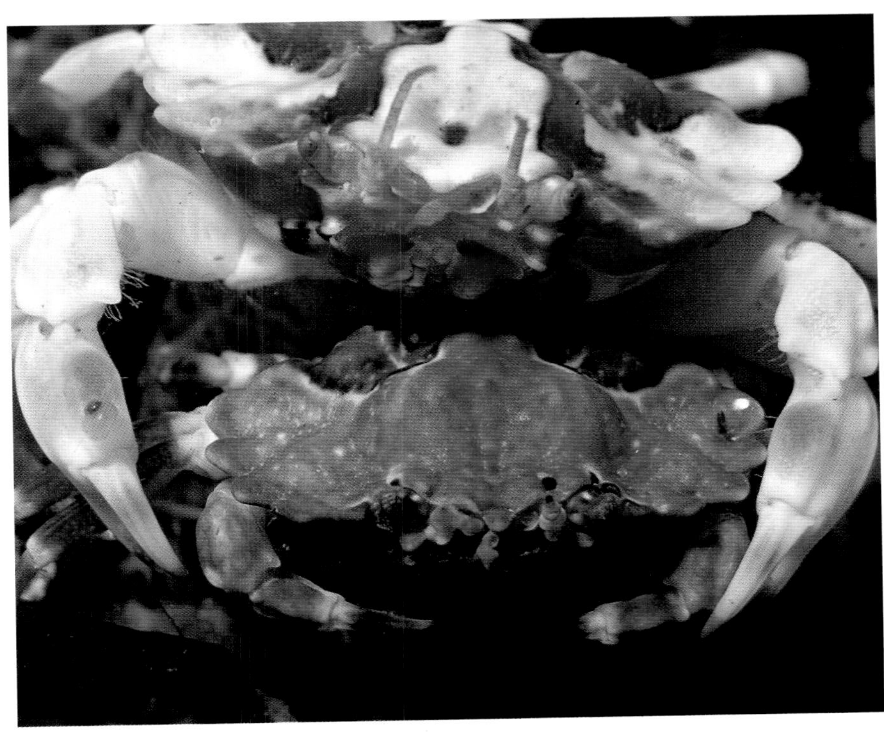

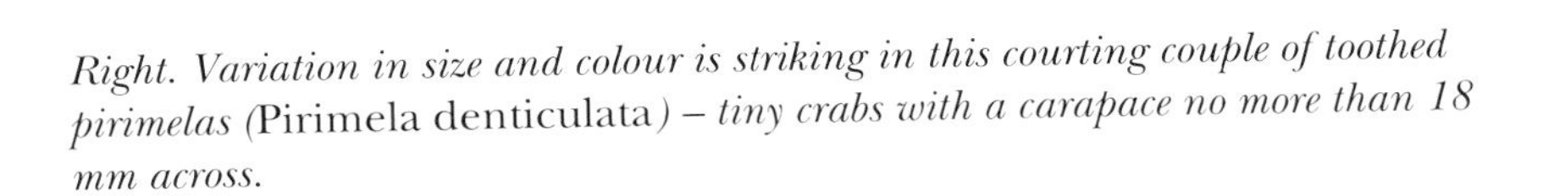

*Right. Variation in size and colour is striking in this courting couple of toothed pirimelas (*Pirimela denticulata*) – tiny crabs with a carapace no more than 18 mm across.*

# CHAPTER FIVE

# WRECKS

For me, the greatest appeal of diving has always been the opportunity to view the beauty and diversity of the marine environment. Anything that does not naturally belong there, including myself, seems something of an intrusion. It was not long, however, before I came to appreciate that some man-made structures can become havens for marine life. Although not a natural part of the sea bed, the metal, wooden or concrete surfaces of shipwreck rarely stay bare for long and may eventually become so heavily overgrown by plants and animals that their original form is totally obscured. Not just wrecks, but piers, harbour walls and other hard structures can all act as artificial reefs and are plentiful in the coastal waters across northern Europe and in North America. Although no animals live on them exclusively, some tend to favour them over other sites as dwelling places. Others that live mainly on rocky reefs may be able to survive in sandy areas by using wrecks as a substitute for rocks. This results in such unexpected finds as jewel anemones off the coast of Sussex, much further east than usual. That stretch of the English Channel offers very few natural hard surfaces, but its many wrecks provide attachment sites or shelter for a variety of animals from depths of a few metres to below 50 metres.

Many of the animals that thrive on wrecks are those that feed by filtering small organisms from the water. The strong currents that often sweep over these sites bring with them plentiful supplies of food particles. Wrecks offer many prominent surfaces, such as masts and other jutting structures, that provide maximum exposure to the flow of water, suiting plumose anemones in particular.

Hydroids and barnacles are some of the first animals to colonise the bare surface of a hard structure. Once these have formed a certain amount of undergrowth, other species, such as sea anemones and large solitary sea squirts, are able to take hold amongst them. Mussels and the soft coral, dead-man's fingers, are also characteristic encrusting colonists. Later, more active animals including sea slugs and bottom-dwelling fish join the community. Wrecks attract many other fish, notably bib and pollack, which are sometimes seen in large schools. The dark recesses of a wreck support little encrusting life, but certain fish find shelter there. Conger eels move into holes in the decaying framework, and large solitary ling may be found in deeper wrecks.

Because wrecks so readily provide beneficial habitats for communities of marine animals and plants, environmentalists have noted their worth and the need to conserve them for this reason as well as for their historic importance.

## *Shoaling fish*

In 1944, a series of floating platforms called Mulberry Harbours was towed across the English Channel to France to be used in the D-Day landings; some sections never made it and remain only a short distance off the English coast. Today, the Outer or Far Mulberry lies two or three kilometres out from Pagham, Sussex, at a depth of about ten metres, and is now so overgrown that its origins are hardly recognisable. With little else but sand for miles around, this wreck, like the smaller Inner Mulberry nearby, is an oasis of life.

One of the most striking features of the Outer Mulberry is the large shoal of bib that is stationed there apparently permanently. Each time I dive there I know I will see the shoal,

*Right. The conger eel (*Conger conger*) remains largely hidden in its lair with only its head poking out to keep watch on passers by.*

*Below. Caught in the beam of my flash, the scales of bib (*Trisopterus luscus*) act like reflectors. The typical, banded pattern is seen in the fish near the bottom of the picture.*

but even after many visits I still experience a sense of surprise when I swim round a corner of the wreck and find myself suddenly face to face with so many fish. Currents can be strong, encouraging the fish to group closely together: row upon row of gleaming coppery flanks displaying pale, narrow bands alternating with broad, darker vertical bands. The chin of each fish sports a distinctive, long white barbel. The shoal forms a dense curtain that parts only when you swim through it. Fish stream to either side and then regroup behind you, all still facing the same way and sheltering against a wall of the wreck in the lee of the current.

Bib (*Trisopterus luscus*) occur all round the British Isles and from southern Scandinavia to the western Mediterranean and Morocco. Although sometimes found on rocky reefs, they are most often encountered around wrecks. Whether rock or wreck is used for shelter, bib like to have sand in the area as well, and so the wrecks of the English Channel suit them perfectly. While younger fish live on shallow sites, older fish occur at depths sometimes well below 50 metres, making it unlikely that they will be seen by sport divers. Bib – also known as pout, pouting or whiting pout – tend to stay in one place when young and may grow to about 30 cm long. They are not generally fished in large numbers, but the family to which they belong, the Gadidae, includes several well-known food fish of great commercial importance, particularly cod, whiting and haddock.

Small shoals of pollack (*Pollachius pollachius*) are commonly seen around wrecks, but the largest and oldest individuals are likely to be found skulking in more sheltered parts of deeper wrecks where they may become well known to a succession of visiting divers. Pollack, like bib, belong to the cod family, though they have a more slender and streamlined shape and can grow much larger, up to about a metre long. They are very similar to saithe, which are of much greater commercial significance, but can be readily distinguished by the lateral line, which is straight and whitish in saithe, and darker and more curved in pollack. Saithe, sometimes known as coalfish or coley, occur both offshore and inshore, whereas pollack have a mainly inshore distribution, from the coasts of northern Scandinavia to the Bay of Biscay.

The pollack's lower jaw lacks a barbel and juts beyond the upper jaw giving these fish a slightly grim air. Smaller fish such as sand eels and sprats are their normal prey, though the young eat crustaceans. The pollack's flanks are often pale, gleaming like silver in the beam of a diver's light, although some have orange spots or criss-crossed markings. Such pigmentation is more pronounced in young fish, and I have seen some individuals so exotically patterned that I thought at first they must be another species. The colour of small pollack living close inshore around rocks ranges from dark bronze to a mixture of pinkish-red markings interspersed with lines of turquoise spots.

## *Sheltering within*

Holes and crevices in wrecks can contain a wide variety of interesting inhabitants, among them fish, and large and small crustaceans including squat lobsters, shrimps and crabs. Perhaps the most spectacular of all is the conger eel (*Conger conger*). The large head and snake-like body, lunging forward inquisitively out of the gloom, is startling. This voracious predator has strong teeth, and many divers are wary of it, although it is dangerous only when provoked.

More heavily built than the common eel (which belongs to a different family), congers grow quickly and may reach a length of two metres or more, though most are considerably smaller. The scaleless skin, covered in mucus, is grey to dull brown with a paler underside. Once, on the *Miown*, a shallow wreck in the English Channel, I was surprised when a fair-sized conger that used to live in the boiler streaked past me in the open across the top of the wreck. Congers usually hide during the day and are not often seen in their entirety, but the fretted structure of a dilapidated wreck may offer glimpses of an elongated body and tail slithering past a series of holes. They may make a wreck their home for several years. I have seen a conger in the same place on the *Soudan*, off Salcombe in Devon, on occasions years apart, though admittedly I could not be absolutely certain that it was the same individual.

Congers hunt at night, catching crabs, octopus, cuttlefish and fish – mainly bottom-dwelling, but also certain open-water species. They have been known to take fish or lobsters from

*Left. This female black-faced blenny (*Tripterygion delaisi*) – much less distinctive than the male of the species – almost matches the encrusting growth of the pier support on which she is sheltering.*

*Facing page. Delicate polyps spread their tentacles atop the tightly packed stems of oaten pipe hydroids (*Tubularia indivisa*). This species has also been recorded in the North American Pacific, but the related pink-mouth hydroid, sometimes called sea strawberries (*T. crocea*), are the most common* Tubularia *there and in the western Atlantic.*

nets and pots, which are often sited around wrecks because these tend to be good fishing grounds. The conger is itself a target of fishing, chiefly for sport but also, particularly in southern Europe, on a semi-industrial basis.

Like common eels, congers have a fascinating life history. They are found from Iceland and southern Scandinavia to West Africa, and throughout the Mediterranean to the Black Sea. After passing through a larval stage, young congers live in shallow water where they favour the supports of piers, among other sites. When adult females are ready to spawn they migrate to their breeding grounds in the warmer waters of the open Atlantic; the exact site probably differs from population to population. In summer, they are reputed to swim to great depths, perhaps 3000 metres or more, and to produce millions of eggs. As soon as this has been accomplished, the females die.

### *Bottom-dwelling fish*

Various small, bottom-dwelling fish including tompot blennies and sea scorpions are very familiar to divers visiting wrecks and other man-made structures, but one species, the black-faced blenny (*Tripterygion delaisi*), has been noticed in northern waters only very recently. This slender fish, which grows to no more than about 7 cm is, like related species, chiefly Mediterranean, though it has also been found in the Atlantic and is now reported frequently by divers in several localities as far north as Dorset, in southern England. Black-faced blennies are members of a family known as three-fin blennies or triplefins, because, unlike true blennies, they have three dorsal fins. Breeding males courting the drably mottled grey or brown females make an arresting sight: their black heads contrast strikingly with the orange or yellowish colour of their bodies, shown off especially effectively when they swim round the females in characteristic figure-of-eight patterns. Sites like breakwaters and the supports of piers are particularly favoured by the fiercely territorial males, each of which may guard the eggs laid by several females.

### *Sea firs*

Certain sea firs are among the first colonists of both shallow and deep wrecks. Squirrel's tail (*Sertularia argentea*) and the extremely similar sea cypress (*Sertularia cupressina*) are hardy and well able to withstand the buffeting from water and sand that strong currents bring to exposed surfaces and to make use of the water flow for filter-feeding. Long, bushy or mat-like colonies of sea firs can be up to 15 cm or more high, usually brownish overall with many fine, white-tipped branches. These sea firs live in a variety of other habitats including stony ground, mussel beds, or the shells of other bivalved molluscs in sand and mud. They are common and widespread in Europe as well as on the Atlantic and Pacific coasts of North America. In certain places they are harvested commercially for decorative uses under the name of white weed.

Sea beard (*Nemertesia antennina*) is a sea fir found in similar habitats and conditions to squirrel's tail, but it establishes itself on wrecks once other species have already settled there, grows more slowly than the squirrel's tail and lives longer. Its pale orange, unbranched main stems are stiff and straight and grow in dense clumps up to 25 cm high.

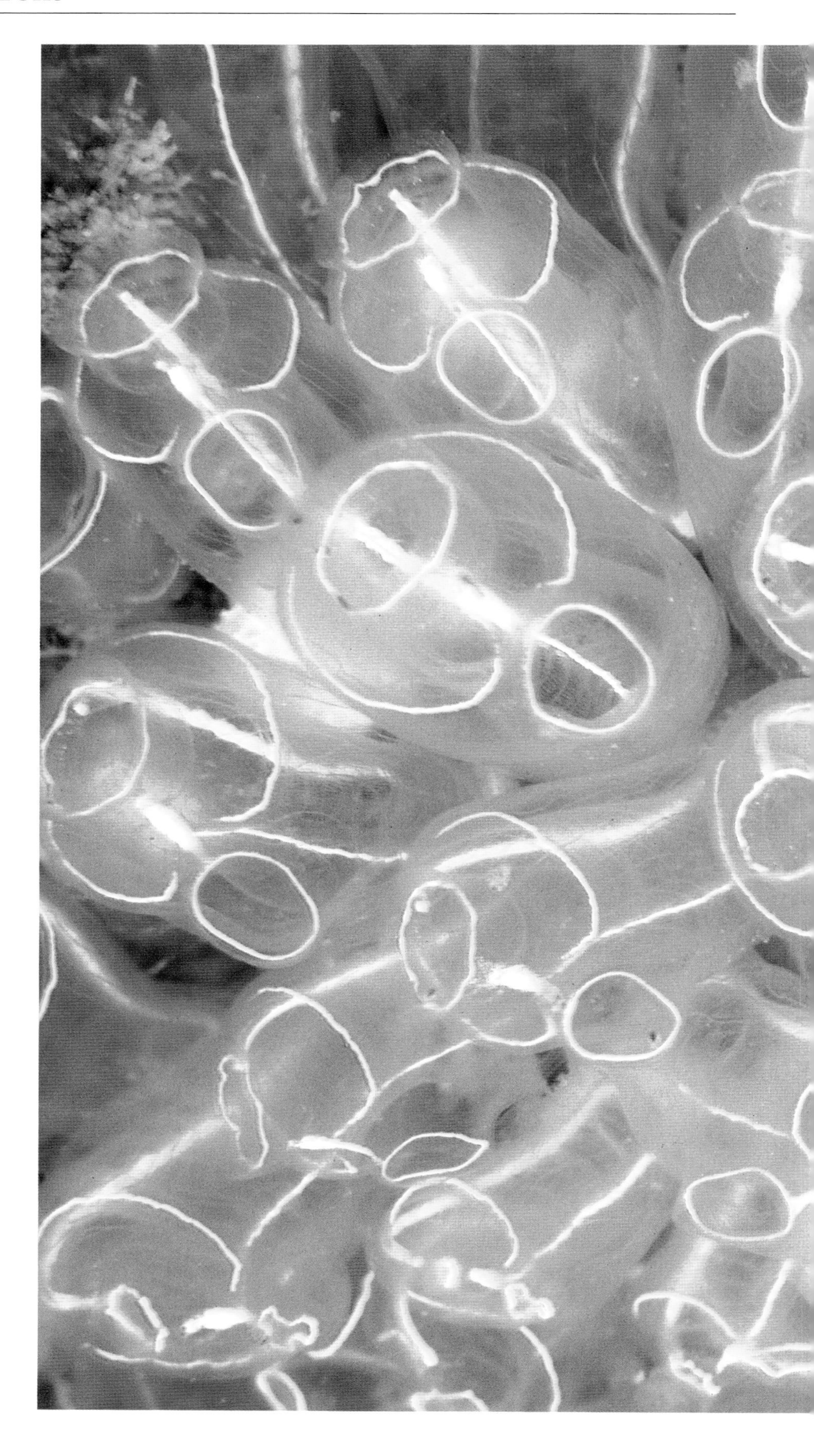

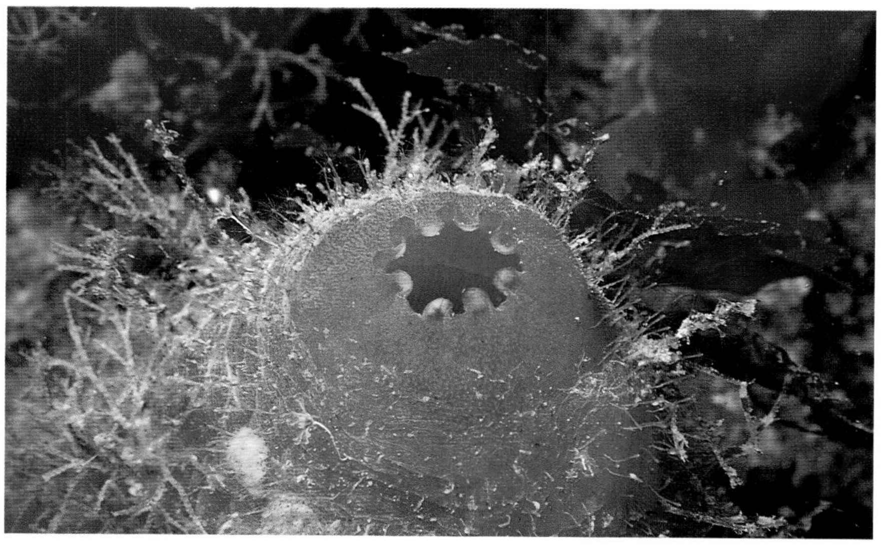

*Above. A dense colony of light-bulb sea squirts, also known as light-bulb tunicates (*Clavelina lepadiformis*). The siphon at the extreme tip is inhalant, while the slightly lower siphon expels the water. Other light-bulb sea squirts occur outside northern Europe, including the North American* C. huntsmani, *from British Columbia to Baja California in the Pacific, and* C. oblonga *in the western Atlantic.*

*Left. Hydroids and bryozoans grow in a fine mesh around the sea squirt* Ascidia mentula. *The animal's oral siphon is clearly visible.*

*Right. The sea squirts* Ciona intestinalis, *here jostling for space on a section of wreck, are fast growing but short lived. This species has a widespread range – including the Atlantic and Pacific coasts of North America, where it is known as the sea vase.* Ciona *has been reported to pump up to four or five gallons of seawater each day to filter its food, obtain oxygen, and excrete waste products.*

*Overleaf. The sea anemones* Sagartia elegans *stand out amongst various other encrusting animals, including small patches of whitish sponges (*Leucoselenia*).*

*Left. A crab's eye view of a plumose anemone (*Metridium senile*) growing on the superstructure of a wreck. The tall column ends in a distinct parapet crowned with a glistening disc which, in this full-grown individual, is divided into lobes. Apart from its wide distribution in Europe, the frilled anemone, as it is known in North America, also lives from the Arctic to Delaware, and from Alaska to southern California.*

*Right. Plumose anemones on a wreck at a depth of about 20 metres. The plain white and orange ones are most frequently seen; the brown-tipped are more rare.*

*Facing page. A particularly beautiful example of the* rosea *colour variant of the sea anemone* Sagartia elegans.

*Below. Cod (*Gadus morhua*) can be seen around wrecks and in various other habitats. The young fish (or codling) hovering here over worm casts on a gravel bed prefers shallow water. This individual is about 16 cm long, adults can reach 100 cm or more. Cod (the subspecies* macrocephalus*) occurs also in North America – on Atlantic coasts from western Greenland to Cape Hatteras, and on Pacific coasts south to California, and across to Japan.*

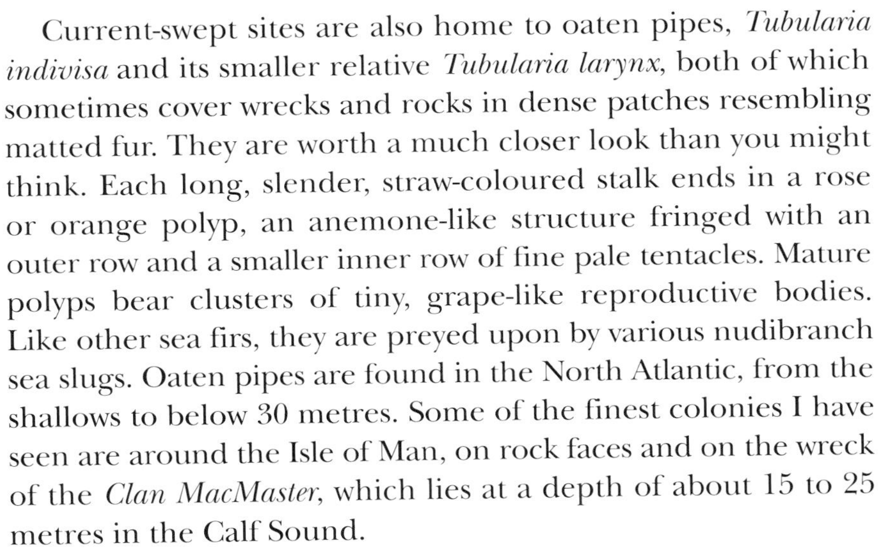

Current-swept sites are also home to oaten pipes, *Tubularia indivisa* and its smaller relative *Tubularia larynx*, both of which sometimes cover wrecks and rocks in dense patches resembling matted fur. They are worth a much closer look than you might think. Each long, slender, straw-coloured stalk ends in a rose or orange polyp, an anemone-like structure fringed with an outer row and a smaller inner row of fine pale tentacles. Mature polyps bear clusters of tiny, grape-like reproductive bodies. Like other sea firs, they are preyed upon by various nudibranch sea slugs. Oaten pipes are found in the North Atlantic, from the shallows to below 30 metres. Some of the finest colonies I have seen are around the Isle of Man, on rock faces and on the wreck of the *Clan MacMaster*, which lies at a depth of about 15 to 25 metres in the Calf Sound.

## *Sea squirts*

Sea squirts are gelatinous or leathery, vase-like structures. Most start life as free-swimming, tadpole-like larvae, but as adults they are permanently attached to rock or some other suitable surface such as a wreck. They feed by filtering organic particles from the stream of water which is drawn in through an inhalant siphon and expelled through the exhalant siphon. These two orifices are close together in the light-bulb sea squirt (*Clavelina lepadiformis*), a transparent species with a white line that runs round the openings and down the side, often likened to the filament in a light bulb. Inside, marking the wall of the pharynx, a mesh of faint lines is just visible. Light-bulb sea squirts are colonial, often forming large rounded clumps in

which the individuals, each about 2 cm high, are joined loosely at the base. Found from Norway to the Mediterranean and the Adriatic, they are common and widespread at all depths down to about 50 metres.

Common sea squirts (*Ciona intestinalis*) are larger than light-bulb sea squirts and grow as separate individuals, not attached to each other although often in groups. The tall cylindrical body may reach a height of about 12 cm but it is soft and able to contract. The outer wall is green-tinged cream to orange, and translucent rather than transparent, revealing narrow, longitudinal bands of muscle and the faint shapes of various internal structures. The rims of the two siphons are large and slightly scalloped, marked with yellow dashes. At the shallow end of its range the common sea squirt often occurs on the supports of piers. Deeper, it lives on many wrecks and sometimes forms dense patches, as on the *Rondo* in the Sound of Mull, West Scotland.

The beautiful pink or vivid red colour typical of *Ascidia mentula* makes it one of the most eye-catching of northern European sea squirts. It has an elongated oval shape, is 5 to 15 cm long, and lives attached by the lower part of one side to wrecks and other hard surfaces. Its top (or oral siphon) is strongly scalloped with white-tipped lobes and is rather more conspicuous than the outlet siphon, which is positioned well down on one side of the body. In its turn, the sea squirt is colonised by

other animals: the surface of its tough outer wall, stippled with minute swellings, provides a suitable attachment site for tiny pale hydroids and sea mats. The range of *Ascidia mentula* extends down to 200 metres; in deep sites individuals tend to be rather more grey and drab.

### *Sea anemones*

Of the various anemones that live on wrecks, the daisy-like *Sagartia elegans* and plumose anemones (*Metridium senile*) are particularly common and abundant. Both are strikingly beautiful, in my opinion rivalled only by the exquisitely coloured jewel anemones (though these are not, of course, true anemones).

*Sagartia elegans*, which I have seen scattered amongst the more drab encrusting life on the sides of a wreck, or covering large areas, is found from Scandinavia to the Mediterranean. It is extremely variable and its similarity to certain other anemones can make identification difficult. Several colour forms have been named, including the exotic-looking pink or magenta *rosea*. Others are pure white, or have an orange disc surrounded by white tentacles, or are a mottled mixture of brown and other colours. They may reach a height of 6 cm and have up to 200 somewhat untidily grouped tentacles, crowning a column dotted with small suckers.

A mass of fine tentacles gives the plumose anemone an almost feathery appearance, but when touched the tentacles instantly withdraw and the long column is reduced to a squat, jelly-like lump. Divers' bubbles have the same effect. Though commonly creamy white in colour, groups of orange and occasionally brown or grey varieties of anemone are sometimes scattered among the white. The typical form – up to 30 cm high – is among the largest of northern European anemones, but they can vary a great deal in size. Plumose anemones also occur along the Atlantic and Pacific coasts of North America, and some of those in the Pacific are immense. In Canada, off Vancouver Island, British Columbia, they reach twice the size of those in Europe.

Wrecks are often colonised by plumose anemones, and after only a few years a dingy hulk can be transformed by a dense carpet of growth benefiting from the nutrients in the water currents flowing over it. Some of the most memorable that I have dived on are the *Robert* off Lundy in the Bristol Channel, the *Lucy* off Skomer Island, West Wales, the *James Egan Layne* off South Cornwall and the *Citrine* off the Isle of Man. The *Lucy* is particularly impressive as it stands upright on the seabed in 35 metres or more – a depth at which divers can spend only a short time admiring the plumose anemones that cover the railings around the decks. However, there are plenty more to see while making the ascent alongside the richly coated mast, which reaches to within ten metres of the surface.

*Above. Tangled seaweeds hang down from a pontoon, becoming entwined with the long fronds of thong weed* (Himanthalia elongata) *growing below.*

*Facing page.*
*Above. A detailed view of dead-man's fingers* (Alcyonium digitatum) *shows the structure of the colony. The individual polyps, resembling slender, translucent anemones, are octocorals, and each has eight, feathery tentacles.* A. digitatum *is also found in the western Atlantic, from the Gulf of St Lawrence to Cape Cod.*

*Below. Dead-man's fingers grow in clumps on a wreck which is also well colonised by hydroids – including oaten pipes – and sea anemones.*

# CHAPTER SIX

# KELP FORESTS

As you swim over kelp, the dense mass of fronds reveals little of what lies beneath, other than the odd, intriguing glimpse as the sea's swell parts the broad ribbons for a moment. A strong swell can give the unsettling sensation that the sea bed seen through the kelp is rushing by while you remain stationary. Not surprisingly, many divers prefer to swim in search of more open areas. Swimming above the 'canopy' of the kelp forest is certainly easier, especially for divers carrying surface marker buoys on lengths of line, which inevitably become entangled. But it is not just the difficulties of swimming through kelp that lead it to being underrated. At sites where kelp grows on the crest of steeper rock slopes, divers often pass it by without a

second glance and head purposefully down to the lower, kelp-free faces. Many tend to regard it as no more than boring vegetation unworthy of exploration. Not so.

Stop and take the time to delve into the kelp jungle, and a wealth of interesting sights will unfold before you. Once you get below the uniform canopy layer, the terrain is varied. There may be a maze of tunnel-like gullies, some connecting with each other or allowing access to small sandy basins while others may be blocked suddenly by boulders. Kelp forests provide shelter for thriving communities of fish and invertebrate animals that live hidden beneath the tangled fronds, among the tall stalks or stipes, or on the spreading holdfasts that anchor the seaweed to the rock. These include sea urchins and tiny cushion stars, sea mats and sea hares. Two-spotted gobies dart about while sea scorpions and dogfish lie in wait, and ballan wrasses hover placidly.

*Below. Small fish swim above the swirling fronds of kelp, swept to and fro by the swell.*

Several different kelps grow on the upper slopes of many rocky reefs in northern Europe, and some of these, including oarweed and sugar kelp, also occur on the Atlantic coast of North America. Sugar kelp (*Laminaria saccharina*) has a short stipe and a broad crinkled frond. Forest kelp (*Laminaria hyperborea*), oarweed (*Laminaria digitata*), and furbelows (*Saccorhiza polyschides*) have palm-like fronds and a long stipe, tubular in the first two but with strongly convoluted edges in furbelows. The holdfast of furbelows is grotesquely swollen and warty, not branched as in the *Laminaria* species.

## *On the fronds*

While swimming over a kelp bed on the way back to shore after a dive, a glint of colour caught my eye among the olive-green fronds. Stopping to inspect it I saw that it was a blue-rayed limpet, tiny but exquisite. Blue-rayed limpets (*Helcion pellucidum*) live singly or in small groups on shallow-growing kelp. The smooth, slightly translucent shell grows up to 2 cm long. Those on kelp fronds are brightly coloured with several rows of gleaming azure dashes radiating from the top of the shell to the margin, whereas those living around the basal holdfast tend to be duller. As the molluscs graze away at the surface of the kelp, they gradually scrape a depression which can seriously weaken the structure of the plant.

A small fish peered round the top of a kelp stipe at me, and I realised it was a baby lumpsucker (*Cyclopterus lumpus*). A careful search revealed several more of these delightful fish, which resemble bright-eyed tadpoles. They attach themselves to the kelp with a ventral sucker but relinquish their hold readily to move round the stipe and fronds, playing hide-and-seek with me as I try to photograph them. The adults are more often found among rocks, but they sometimes lay their eggs on kelp, and young fish are likely to be seen on that or other seaweeds.

Many other small animals live on kelp fronds, some anchored, others mobile, feeding on the seaweed itself or preying on other occupants. Jelly-like lumps on the fronds, squat and somewhat angularly bottle-shaped, are likely to be the gas-mantle ascidian (*Corella parallelogramma*), which grows to a height of about 5 cm. Its transparent outer coat, flecked with white or yellow, reveals all the internal workings: a mesh is the branchial sac or filter-feeding pharynx, a curved line is the whitish gut, and a basal mass which is often orange, the gonad.

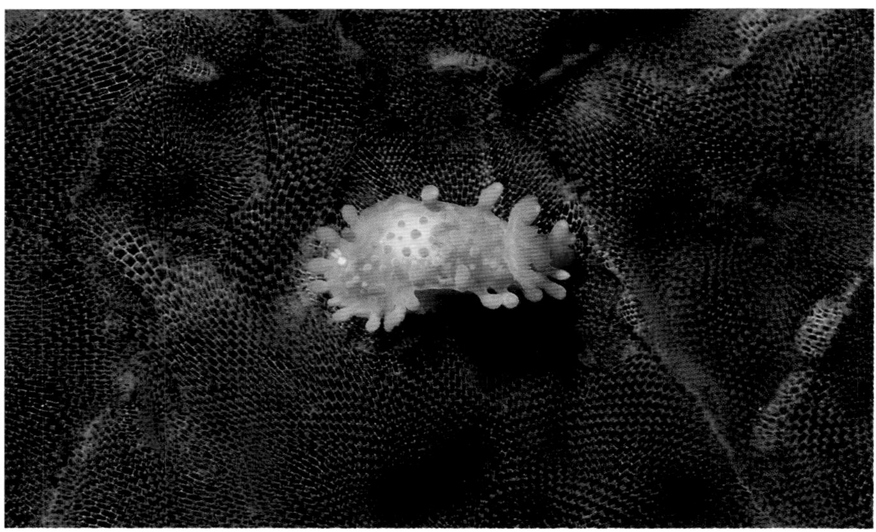

*This page.*
*Nudibranch sea slugs,* Onchidoris muricata *(above), which grows up to 14 mm long, and* Limacia clavigera *(left), both feeding on the sea mat* Membranipora membranacea. O. muricata *is common in northern European waters and also occurs on both coasts of North America.*

*Below, left.* Polycera quadrilineata, *creeping past several tiny gastropod molluscs on the edge of a kelp frond, is a mere 15 mm long, but may grow to twice that size; right, a blue-rayed limpet (*Helcion pellucidum*) grazing on kelp, with a minute individual barely 2 mm long among small patches of sea mat to the right in the picture. Juvenile limpets more usually settle on other, comparatively tender, brown seaweeds (particularly wracks), until they are about 4 mm long and able to eat the tough kelp fronds; this limpet seems to have graduated early.*

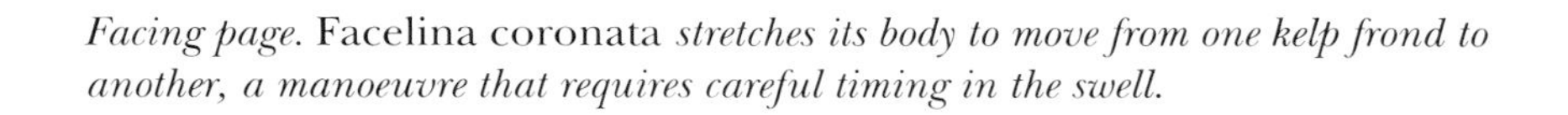

*Facing page.* Facelina coronata *stretches its body to move from one kelp frond to another, a manoeuvre that requires careful timing in the swell.*

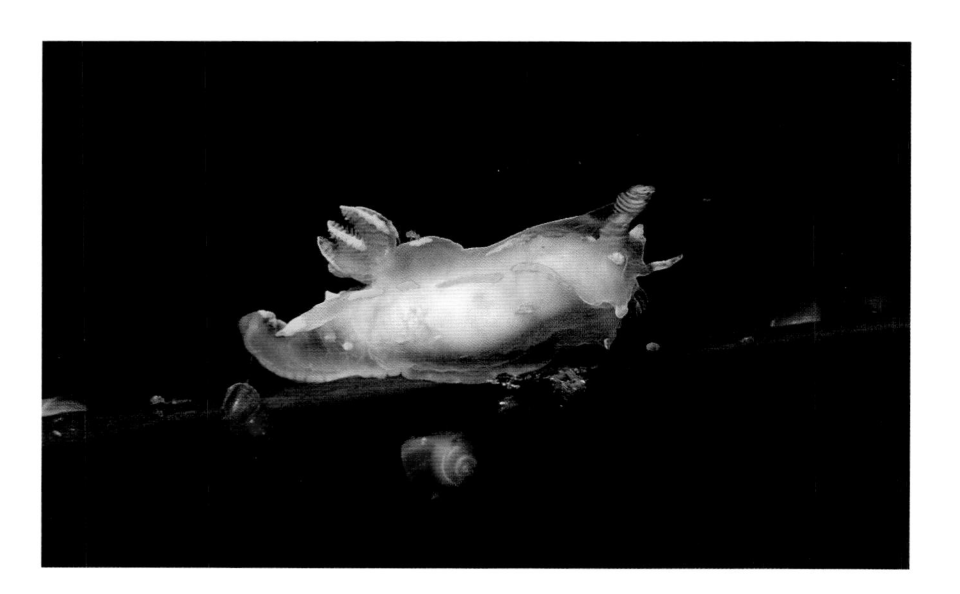

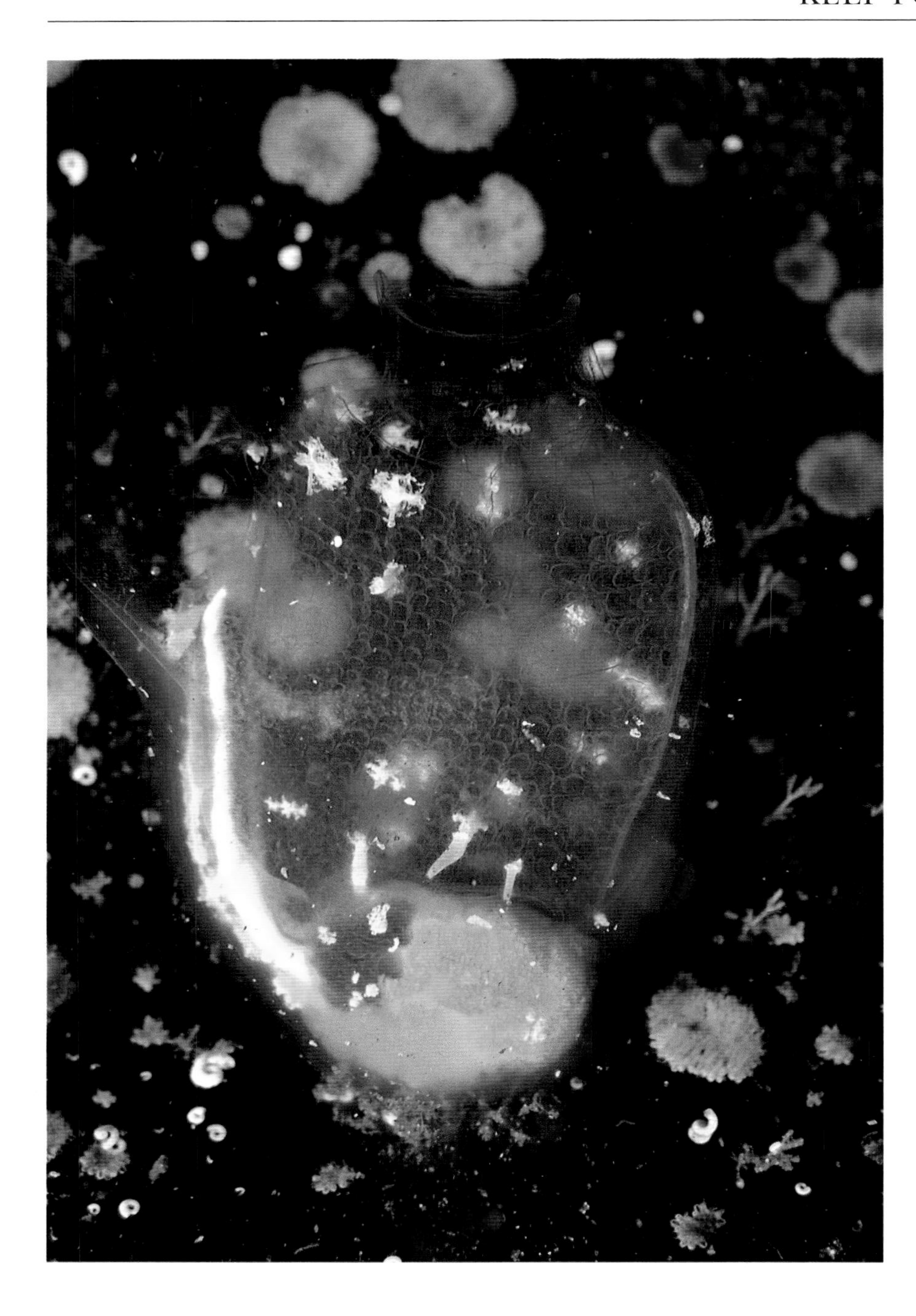

*Left. The orange gonad and other organs show clearly through the transparent walls of this gas-mantle ascidian* (Corella parallelogramma) *attached to a kelp frond. The top opening of the sea squirt is the intake siphon, and the spout-like siphon is the outlet. Related species occur on the Pacific coast of North America:* C. willmeriana *from Alaska to California, and* C. inflata *in Puget Sound.*

This ascidian, or sea squirt, sometimes attaches itself to other, firmer structures such as stone or shell.

The fronds and stalks or stipes of *Laminaria* are often overlaid with whitish patches several centimetres across which have a delicate lace-like structure. This fine mesh is extremely regular, made up of minute rectangular compartments with a blunt spine at each corner, and the wavy border of the whole patch has a slightly denser texture. At first glance the patch might appear to be merely a pattern on the plant, but, amazing though it may seem, it is a complex animal: the sea mat *Membranipora membranacea.* Within each rectangle it is just possible for a keen eye to make out the ring of waving tentacles extended by the individual animals in the colony to trap food particles. Sea mats are fast-growing colonisers and are widespread in temperate seas. Their North American relatives are known as lacy crusts.

Sea mats are eaten by some sea slugs, including *Onchidoris muricata* and *Limacia clavigera.* Both are small, white animals, the former covered with tubercles and the latter handsomely decorated with yellow or orange raised spots, finger-like projections, and yellow-tipped rhinophores (sensory tentacles) and gills. *Limacia* feeds on various encrusting bryozoans, particularly the hairy sea mat (*Electra pilosa*), which is common on toothed wrack and certain other algae and also on rocks and stones. In northern Europe, *Limacia clavigera* grows to about

*Below. Finding no bare space to spread over as it grows, a colony of the star sea squirt (*Botryllus schlosseri*) forms a layer on rock already encrusted with coralline algae and barnacles. The sea squirt is also known in the western Atlantic (Bay of Fundy to North Carolina).*

18 mm long, but off western Africa it may grow even bigger. Another sea slug found eating sea mats, *Polycera quadrilineata*, is typically also white with yellow markings but has a rarer form that is dark speckled. Most beautiful of all is *Facelina coronata*, shimmering with blue iridescence as it feeds on tiny hydroids on the kelp fronds.

## *Around the base*

Like miniature patchwork quilts cast over kelp holdfasts and rocky surfaces among the kelp, the star sea squirt (*Botryllus schlosseri*) has an embellishing effect on its surroundings. Again, the intricate patterns indicate the arrangement of the individual animals that make up the colony. Each 'star' is in fact a

*Above. The cushion star (*Asterina gibbosa*) on a kelp frond with small crusts of sea mat (*Membranipora membranacea*). A much larger North American relative of the starfish, the bat star (*A. miniata*), is found on rocks and sand in the Pacific from Alaska to California. The sea mat occurs there too; other* Membranipora *species live on western Atlantic coasts.*

*Left. Colonies of the same sea mat carpet a kelp frond, where a painted top shell (*Calliostoma zizyphinum*) browses on small hydroids and algal growth. Other top shells include the rarer* C. occidentale, *known from northern Europe and the Atlantic coast of North America.*

*Facing page. The spiny starfish (*Marthasterias glacialis*) gripping a stipe of the kelp,* Laminaria hyperborea, *which has a characteristic growth of red seaweeds on it.*

*A dogfish* (Scyliorhinus) *resting against an algal-coated rock. Minute, pale bluish specks behind its head and around its gill slits are parasites, probably the young stage (known as praniza larva) of isopods. These jump off if startled, and when adult leave their host permanently to take up a free-living existence.* Scyliorhinus *species occur in Europe and in the North American Atlantic.*

*Right. A two-spotted goby* (Gobiusculus flavescens) *hovers above the sea bed, displaying its colourful marking.*

group of individuals around a common exhalant siphon. Many colour variants exist, the 'star' usually contrasting with the background.

The spiny starfish (*Marthasterias glacialis*) is a common inhabitant of the walls of kelp gullies, sometimes clinging to the kelp itself by wrapping its five, long, pale grey arms tightly around the thick lower end of the stipe. Prominent whitish or mauve spines are arranged in regular rows along its arms, and it is sometimes tinged with green or brownish yellow. Each major spine is set at the centre of what to the naked eye looks like a small pincushion but is actually a cluster of tiny, pincer-like appendages, the pedicellariae, which are used by starfish to clean themselves of debris. The spiny starfish is fairly abundant on various types of sea bed, particularly rock or stony ground around the western and south-western coasts of Britain, and occurs north to Norway. It commonly measures about 25 cm across and occasionally grows much larger on a varied diet that includes other echinoderms, such as smaller starfish, as well as crustaceans and molluscs. Like the common starfish, it uses the powerful grip of its arms to prise open the shells of bivalve molluscs so that it can feed on the fleshy animal inside.

Here, too, among the kelp, lives a much smaller starfish. The cushion star or starlet (*Asterina gibbosa*) has the short-armed, fat-bodied shape typical of other cushion stars but, unlike most of them, it grows to no more than 5 cm in diameter and often only reaches half that size. Its stiff, greyish-green or pale brown upper surface is covered in small clusters of short, orange or yellowish spines. Common from southern and western British coasts to the Mediterranean, it feeds on organic detritus on the sea bed and generally lives in the shallows, sheltering beneath overhangs and stones, though it can also be found deeper.

As grazers on small seaweeds, painted top shells (*Calliostoma zizyphinum*) find plentiful supplies of food on the rocks at the base of the kelp and attached to its stipes. The sharply pointed, conical shell, up to about 30 mm high, has a ridged outline of spiral whorls, but its purplish-pink colour and chequered pattern may be muted by a fine layer of algae or even barnacles. Unless the mollusc is disturbed, its cream or yellowish head and foot can be seen protruding from the shell. A pair of long, slender tentacles reach out to probe the surroundings, with a tiny black eye on a stalk at the base of each. The

top shell varies its algal diet with the polyps of hydroids and dead-man's fingers. It lays pale yellow eggs on stones or seaweed, where I have occasionally found the coiled band of the egg mass with the mollusc still beside it. The painted top shell is not restricted to the shallows, and ranges down to 300 metres around British and Irish rocky coasts and from Norway to the Canary Islands, the Azores and the Mediterranean.

Common sea urchins (*Echinus esculentus*) are easily dislodged from rock faces, so they tend to seek out ledges, crevices and gullies where they are unlikely to be swept away. Kelp beds offer ideal protection to these fragile animals, which are encased in a brittle shell (or test) and densely covered in equally brittle spines. They also provide food in the form of both the kelp itself and other rock-encrusting algae and animals living there. Seen in the shallows, these urchins are pink, but on deeper parts of the reef they look bluish because water absorbs the red end of the spectrum more strongly. Membranous tube feet can be extended well beyond the spines and, despite their delicate appearance, enable the animal to move. This species occurs from Iceland to Portugal, but is absent from the Mediterranean. It is abundant in some parts of its range, but rare in others and may be at risk of over-exploitation by souvenir hunters. Sometimes referred to as edible sea urchins, or even – in the past – as sea eggs (because they were dipped for a

*Facing page. In European waters, kelp grows in dense stands on shallow rock, although it never reaches the towering heights of its North American counterparts on Pacific coasts. The encrusted stipes are typical of forest kelp (*Laminaria hyperborea*). The animals living on the sheltered walls of the gully beneath include the starfish* Asterias rubens, *brittle stars and sponges.*

*Below. The long, elastic, tube feet of the sea urchin* Echinus esculentus *reach out in all directions to sense the environment, as the urchin moves slowly over the sea bed.*

*The heavy-bodied ballan wrasse (*Labrus bergylta*) patrols its territory unhurriedly. By night, it hides in the weeds or tucks itself into a crevice.*

short time in boiling sea water and then eaten like eggs) all the contents, except the gut, are edible, but more often only the roe is scooped out.

## *In the gullies*

The enjoyment of swimming along a gully lies in the possibilities, as yet unseen, hidden behind the next boulder or twist in the trail. Dogfish (*Scyliorhinus*) are not uncommon, but I never fail to feel a surge of excitement when I see one, perhaps because they are larger than most fish in northern seas or else because they are unmistakably sharks. Like other sharks, they are cartilaginous and have no true scales; instead the skin is as rough as sandpaper. Dogfish are easily approached by day, when they conserve their energy for their nocturnal hunting activities. They move from their resting places on the sea bed only when they judge it necessary, and then swim away with sluggish reluctance, body twisting from side to side. Their translucent egg capsules or 'mermaid's purses' can be found twined around seaweeds. In Britain, dogfish is sometimes marketed under the more appetising-sounding name of rock salmon.

In the quiet backwaters of kelp gullies, groups of tiny fish flit about either in the open or seeking tiny crustaceans among the weeds. Two-spotted gobies (*Gobiusculus flavescens*) have a different lifestyle from most other gobies and, instead of resting on the sea bed, hover for most of the time in mid water. The slender, orange or reddish-brown body has a black spot at the base of the tail fin and, in males, another spot part way along each side. Cream or yellow streaks across the tail and one or more rows of pale blue spots along the body make this goby conspicuous when it is in the open, despite its small size of 6 cm or less. It often lays its eggs in hollow holdfasts of furbelows. It can also be found on other shallow sites, in eel grass (*Zostera*) beds and occasionally in rock pools, from Norway to north-western Spain.

By contrast, ballan wrasses (*Labrus bergylta*) appear massive as they patrol the kelp gullies and other shallow rocky sites, singly or in scattered groups. At up to about 50 cm in length, this is the largest European wrasse, heavily built and with thick lips. Strong teeth in the jaws are aided by pharyngeal teeth

*Right. A young lumpsucker (*Cyclopterus lumpus*) perching beside small gastropod molluscs on a kelp frond, secured by a ventral sucker. This small fish, barely 4 cm long, may grow to 50 cm or more. Lumpsuckers occur widely in northern Europe and in the North American Atlantic, in Greenland and south to Maryland.*

*The sea hare (*Aplysia punctata*) crawling over branching fronds of shallow-growing seaweed. Other sea hares are found on both coasts of North America –* A. willcoxi, *for instance, occurs in the Atlantic from Cape Cod to the Gulf of Mexico.*

behind the gills to crush the mussels and crabs that form part of its diet. Ballan wrasses are slow growing and may live for 20 years. They all start out as females, and it is several years before some eventually become sexually mature males. Young fish tend to be greenish and old males more red, but the typical colour is orange to reddish-brown, often with a distinct, speckled or chequered appearance given by the pale green- or whitish-centred scales. Ballan wrasses occur from southern Norway to the Canary Islands and North Africa, but are rare in the Mediterranean. Along with certain other wrasses they are also sometimes seen on wrecks.

## *Red seaweeds and grazers*

Various red algae, including the beautiful, iridescent *Drachiella spectabilis*, grow on the bedrock or on kelp stipes. Here and on other rocky sites, sea hares (*Aplysia punctata*) graze, and in early summer groups can be seen clambering over one another to mate. Sea hares are sea slugs, but unlike the nudibranchs they have a shell, albeit a tiny, fragile, internal one. Spawning usually takes place on shallow-growing algae, and sea hares migrating to an area of green seaweed are reported to change from their typical reddish-brown colour to an olivey green. The strings of eggs are laid in tangled, pinkish or orange coils. When they hatch out, the larvae, unlike the bottom-living adults, are covered by a shell and swim in open water.

*Facing page. The tube worm* Bispira volutacornis *extends its large spirals of tentacles (gills) for filter feeding. These are often whitish or buff, but they are sometimes orange, or, as here, banded with darker brown.*

# CHAPTER SEVEN

# HIDING IN CREVICES

Rocky sites and wrecks are home to a far more extensive and varied population than one might at first suspect, for many animals prefer the safety of holes and crevices to more exposed situations. Even a sheer rock face has cracks and fissures that make excellent hiding places, and the gaps between boulders and stones, and underneath them, provide further refuges. The boldest among crevice-dwelling fish and crustaceans usually keep a look-out from the safety of their shelters, lurking at the entrances. To make out what lies further back in the gloom, a small torch is useful and will reveal in its beam the colours or beady eyes of the squat lobsters, prawns and other creatures found there.

The more active animals – crabs, lobsters and octopus – venture out into the open some of the time, particularly at night when they hunt for food, using their retreat to rest in by day. Some others, including tube worms and certain sea cucumbers, are sedentary and remain tucked well into a crack, with only their gills waving about outside to capture food particles from the currents. If you spot sturdy arms snaking from a crack in the rock, they may well belong to the crevice brittle star (*Ophiopholis aculeata*). Standing out from the shadowy recesses between boulders, the pale feathery whorls that are the tentacles (or gills) of the tube worm *Bispira volutacornis* (a southern species common from the Mediterranean to the Isle of Man) are easily identified but not so simple to approach. Like sand-dwelling peacock worms, they are very sensitive to sudden movements nearby, detecting changes in the water current and withdrawing rapidly into their tubes, often foiling any attempt to photograph them. What you usually end up looking at close to are just their membranous tubes which are grey, rubbery and mud-encrusted.

Various fish live in holes in rocks, particularly blennies and gobies, and, of course, conger eels, which are so at home in wrecks. Rocklings, including the three-bearded *Gaidropsarus*

*vulgaris*, an elongated fish that sports a barbel on its chin and one on each nostril, also hide among rocks as adults, although, as in many fish, their juveniles drift in open water. Wrasses hover around rocky areas and use crevices as nesting and spawning sites.

## *Lobsters' lairs*

A pair of massive front claws bars the entrance to the cave of the fiercely territorial common lobster (*Homarus gammarus*). The claws are formidable weapons both for crushing prey and for fighting off unwelcome intruders; they require considerable muscle power to lift, let alone to wield, and should be treated with respect. In life, the lobster's carapace is a surprisingly deep purple or blue, but cooking destroys its blue pigment, leaving just the red and producing the 'boiled lobster' appearance. Found in both shallow and deeper water – there is some evidence that they may head deeper in winter – lobsters occur from Scandinavia to Morocco and the western Mediterranean. Seen as a luxury food almost everywhere, they are caught in such large numbers that they are probably over-exploited.

Peering into the cracks between the rocks, I was startled by a flash of red and blue, and then identified the brilliantly coloured body of a much smaller crustacean, the spiny squat lobster (*Galathea strigosa*), its long legs appearing almost furry from their thick covering of bristles. I waited patiently, willing the animal to venture far enough forwards to allow me to

*Above. The common prawn (*Palaemon serratus*) is transparent in life apart from thin, brown stripes and yellow dashes. It hides under ledges or in weed and is often found in rock pools and other shallow sites. Several other* Palaemon *species live in Europe; elsewhere they include* P. macrodactylus, *an oriental shrimp introduced to California and South America.*

*Left. A common lobster (*Homarus gammarus*) – about 30 cm long (it could grow up to 50 cm) – thrusts its huge claws forwards defensively. At night, it uses them to tear its prey apart, crushing with its right claw and cutting with its left. Its North American counterpart is the northern lobster (*H. americanus*) on the Atlantic coast from Labrador to Virginia.*

*Facing page. A strikingly coloured spiny squat lobster (*Galathea strigosa*) has left its narrow crevice at night, in search of food. In squat lobsters the abdomen is curled under, and they are more closely related to hermit crabs than to straight-tailed true lobsters.*

photograph it, while trying to steady myself against the sea's uncomfortable swell. Eventually my dive time ran out and I was forced to leave, frustrated by my wasted efforts. Disappointment was soon forgotten though, when, diving again after dark, we found several squat lobsters sitting brazenly in the open, preparing to scavenge scraps off any dead animals they could find. At up to 12 cm, the spiny squat lobster is the largest bodied of the several species found in northern Europe and occurs from Scandinavia to the Red Sea.

A recess under an overhang may be spacious enough to house several common prawns (*Palaemon serratus*) along with larger crustaceans, provided they each keep to their own corner. A smaller, more snug crevice is likely to be occupied by a single animal, like a lone crab, that can wedge itself against the walls if besieged. Edible crabs (*Cancer pagurus*) spend much of their time in cracks in rocks or under boulders; on softer ground they are able to create their own shelter by excavating a hole. A favourable patch may be occupied by large numbers. One such is the steep-sided Mixon Hole, off the Sussex coast, southern England, where the flat sea bed drops away suddenly from about 5 metres to more than 25 metres. Making a descent alongside this face is like taking a ride in a lift down the outside of a high-rise block. Narrow ledges divide the soft, clay wall into storeys, which are pitted with holes. In nearly every hole sits a crab, large or small, with solid, black-tipped claws folded across its front and small, green eyes impassively surveying the world from the shelter.

Edible crabs are found from the shore down to about 100 metres, the largest, thick-shelled and ponderous, usually at the deeper sites. The heaviest on record weighed over 6 kg, but even a 2 kg crab that I encountered on a wreck in the English Channel was an impressive sight. The elongated oval carapace may grow to about 25 cm across. Regular indentations around the front edge give the distinct impression of a piecrust, heightened by the pinkish or orange-brown colouring. Young crabs are smooth shelled, but older crabs often become encrusted with barnacles or the small, curled tubes of keel worms (*Pomatoceros*). Edible crabs are found from northern Norway to the Mediterranean and West Africa, and crab fishing is an important industry. Some countries have set a minimum legal size at which crabs can be caught in order to protect young crabs and to allow populations to replenish themselves.

The velvet swimming crab (*Liocarcinus puber*), a more lightweight animal, is skittish and often seen out in the open. When approached, it makes an aggressive stand with claws outstretched, only retreating to a crevice if this fails to impress.

### *Gobies and blennies*

Gobies and blennies are typically bottom-living fish, found, on the whole, in crevices or under stones. The rock goby (*Gobius paganellus*) is one of the larger species, sturdily built and up to 12 cm long, but easily missed unless you peer under small boulders. It is a loner, waiting to catch the crustaceans up to small crab size that live in the shallows and in rock pools. Unusually for a goby, it can live up to ten years. Smaller species such as the two-spotted goby rarely live longer than two years, an age at

*Facing page. Weighed down by its armour, an edible crab (*Cancer pagurus*) rests its large claws on the clay ledge in front of its crevice. Related crabs include several North American species, such as the red rock crab (*C. productus*) in the Pacific, and the Atlantic rock crab (*C. irroratus*).*

which the rock goby has only just reached sexual maturity. Another shy species, the conspicuously blotched leopard-spotted goby (*Thorogobius ephippiatus*), lives in crevices at the foot of underwater cliffs, where you will probably see it sitting near the entrance ready to retreat into the darker recesses. Leopard-spotted and rock gobies are found from Ireland, southern and western Britain to the Mediterranean and the Canary Islands, and the rock goby is also known in North Africa and the Black Sea.

Timidity might seem prudent for a small fish encountering a diver, but the natural curiosity of a tompot blenny (*Parablennius gattorugine*) usually outweighs its fear. When I come face to face with a tompot, its first reaction is to retreat a little into the safety of a hole, but then if I wait patiently for half a minute or so the small head pops out to see what is going on outside. So long as there is no sudden movement, the blenny will normally venture close enough to allow me to take its portrait. The thick lips are usually slightly parted and the bushy reddish tentacle above each eye adds to its air of wide-eyed surprise. A tiny tentacle also sprouts from the edge of each nostril. The scaleless body is somewhat elongated and may be up to about 20 cm long, though typically considerably smaller.

Tompot blennies occur from Scotland to the Mediterranean and Morocco. The male takes up residence in a small hole in rock or in a shallow wreck, from which he peeps out while

*Below. A velvet swimming crab (*Liocarcinus puber*), at about 8 cm across the carapace close to its maximum size, resting on a bed of the anemone-like* Parazoanthus anguicomus. *Paddle-shaped hind legs enable the crab to swim a few strokes as it scampers from rock to rock.*

*This page, right, from top to bottom.*
*A rock goby (*Gobius paganellus*) peers warily from the entrance of the hole it inhabits in a steep, clay wall.*

*Iridescence sparkles on the scales of the leopard-spotted goby (*Thorogobius ephippiatus*) resting beside sponges. The warty yellow sponge is the orange puffball sponge (*Tethya aurantium*), another variety of which occurs in the North American Pacific, from Alaska to Baja California.*

*A tompot blenny (*Parablennius gattorugine*) makes a defiant stand for a small fish, revealing minute but sharp teeth between its parted lips.*

*A goldsinny (*Ctenolabrus rupestris*) advances with rapt attention on a purse sponge (*Scypha ciliata*) among red seaweed. The wrasse's small teeth are bared in readiness to pick off any minute invertebrate it spies.*

*Facing page. A perky tompot blenny (*Parablennius gattorugine*), with large red tentacles, eyes me inquisitively. Similar species occur in southern Europe and in the North American Atlantic, including the seaweed blenny (*P. marmoreus*) from New York to Venezuela.*

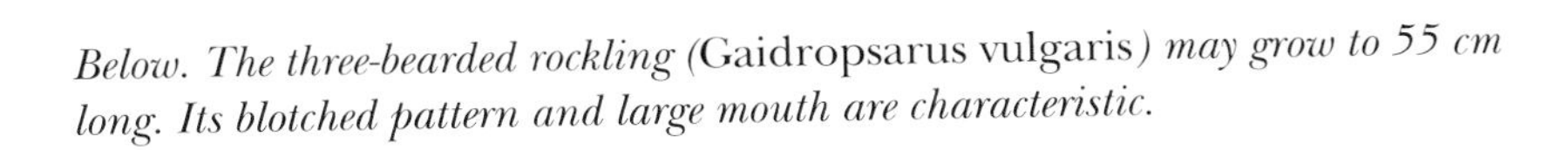

*Below. The three-bearded rockling (*Gaidropsarus vulgaris*) may grow to 55 cm long. Its blotched pattern and large mouth are characteristic.*

guarding eggs laid there by females. Some of the larger, dark red and brown males are remarkably bold, more inclined to rush out at me than to hide, as if I might be a rival to be driven away. When I meet such a feisty blenny, I settle down in front of it, a couple of feet away, and brush the top of a nearby pebble with my finger. A few specks of debris drift up through the water, sufficient enticement for the blenny to dart forward eagerly to see if I have uncovered a morsel worth eating. The blenny may come right up to my camera lens so that I have to back away. From then on the fish follows me as I back off, perhaps sensing that it has the upper hand. Behaviour, and colour too, put me in mind of the robin that perches near me when I am gardening.

It is hard at first to imagine any fish less like a blenny than that ugly giant, the wolf-fish (*Anarhichas lupus*), but this is a relative, albeit a fairly distant one. The wolf-fish or catfish has a large rounded head with a blunt snout and a large mouth. Its prominent teeth are heavy and irregular, well suited to crushing crabs, molluscs and sea urchins; not surprisingly, the teeth are quickly worn down and – fortunately for the wolf-fish – replaced each year. The long body extends to more than a metre, with wrinkled skin that is typically bluish grey to brown. Young wolf-fish may be more brightly coloured, even orange or red, and I have sometimes initially mistaken small ones for an unusual kind of blenny. To me the wolf-fish's ugliness has a certain charm, but to some a large adult might appear rather fearsome. However, they are not ferocious unless severely provoked: one account tells of a descending diver landing heavily on one, which caused the alarmed fish – not unreasonably – to bite.

Wolf-fish are known at depths well below 50 metres, but some live in much shallower water where they are more likely to be seen by divers. Individual fish may inhabit a particular cave for a long time and, as at Eyemouth in Berwickshire, Scotland, for instance, have become well known, receiving regular visits from divers. Wolf-fish have a northern distribution and are fairly common from Norway and Iceland to southern Scotland. They also occur off the Atlantic coast of North America from Labrador to Cape Cod and Greenland, whereas a similar fish, the shorter bodied wolf-eel (*Anarrhichthys ocellatus*), is found only on the Pacific coast of North America. They are fished in some places, but are of minor commercial importance.

### *Wrasses*

Wrasses often hover beneath overhangs and ballan wrasses sometimes rest at night in crevices. In summer, the male builds a nest in a crevice, where the female will lay her eggs. The smaller goldsinny (*Ctenolabrus rupestris*), a wrasse that may reach a length of 18 cm but is usually much shorter, is even more likely to be seen in such haunts. Unusually among wrasses, the sexes look alike, with large scales and a colour range from golden or bright orange to reddish-brown, contrasting with the whitish underside and the distinctive black blotch on the back by the tail fin. When you come face to face with this fish, its sharp front teeth are obvious, protruding slightly from the small mouth. It may give you a nip, but it is more likely to turn its attentions to edible morsels on the rocks. The goldsinny's diet, mainly of small crustaceans and worms, includes parasites picked off larger fish – behaviour shared with tropical members of the family, the small cleaner wrasses. It has now become standard practice in commercial salmon farming to include a number of wrasses as cleaners, either goldsinnies or other species that act in the same way. Goldsinnies live alone or in small groups, mainly among rocks and weed, but occasionally in sandy, sea-grass beds, from southern Norway to the Mediterranean and Morocco.

*Top. A wolf-fish (*Anarhichas lupus*), with the looks of a veteran prize-fighter, watches from its cave beneath a ceiling covered with the tubes of worms. Remnants of a crab consumed earlier lie discarded in the foreground.*

*Left. The lesser octopus (*Eledone cirrhosa*) hunts by night on open sand, but during the day it hides among rocks to escape predation.*

# CHAPTER EIGHT

# DEEPER REEFS

In murky northern seas, descending through the shallows to a deeper sea bed is very different from doing so in the clear, sunlit tropics. Looking down, the column of water appears endlessly dark and impenetrable, while above, daylight rapidly dims as the surface layers close over your head. The chill, real and psychological, increases with a descent that seems to go on for ever until at last the sea bed looms up. The depth may be no more than 20 to 30 metres, but it can feel as if the surface is light years away. The deeper you go, the less time you can spend without risk of decompression sickness; a dive like this needs a certain amount of experience and more careful planning than usual.

The rewards are there, though. On reaching the bottom, or breaking a free fall down an underwater cliff, inspection of the terrain reveals a significantly different assortment of inhabitants to those of the shallows. These are deep-water species, not in the strict sense but in terms of what a sport diver is likely to encounter. In the reduced light levels, the kelp peters out, and other algae grow only sparsely. Many encrusting animals may occur on the rocks, some low-growing but others large and prominent. The more sheltered conditions away from wave action or swell allow fragile growths such as ross corals and sea fans to flourish. These and various other exotic-looking species appear incongruous in their cold northern setting, as if transplanted from a tropical coral reef.

*Pages 108-109. A colony of* Parazoanthus axinellae *sprouts from a sponge-covered rock face beneath an overhang.*

*Pages 110-111. A colony of* Parazoanthus anguicomus, *known at depths as great as 400 metres, is a more northerly species than* P. axinellae, *which may confusingly also be whitish but has fewer tentacles.*

*Below. The magnificent crawfish,* Palinurus elephas, *has a pair of strong antennae that may extend to double the length of its body, which itself can be up to 50 cm.*

The firm but springy branches of the orange or brownish sponge *Axinella dissimilis* grow like long, twisted and knotted fingers, usually on upward-facing rock on deeper reefs. Colonies of the anemone-like *Parazoanthus axinellae* are often found living on relatives of *Axinella dissimilis*, although they sometimes encrust more solid substrates, including shells and worm tubes. These species, living in association, have a southern distribution and are prevalent in the Mediterranean, but they can be seen at a few more northerly sites. The yellow polyps of *P. axinellae* grow up to 15 mm high and have about 30 slender tentacles, whereas a closely related species, *P. anguicomus*, is white and has more tentacles, usually about 40.

A species of true sea anemone, *Bolocera tuediae*, with its delicate hue, graceful form and impressive size (up to 30 cm across the tentacles) ranks as one of the most beautiful in northern Europe. It resembles a dahlia anemone, but is a uniform pale pink, creamy orange or almost white and has a smooth column topped with up to 200 long, thick tentacles. It lives attached to boulders or shells, and I have seen it on brittle-star beds at around 20 to 30 metres. It has also been found at depths as great as 2000 metres and has a northerly distribution extending to the Arctic Circle and also to the east coast of North America; it is rare south of Scotland.

Deeper living animals also include rounded colonies of the football sea squirt (*Diazona violacea*), which is similar to the more shallow-living light-bulb sea squirt but less transparent, encased in a common test or tunic with a translucent greenish hue.

Long-bodied ling (*Molva molva*), fished commercially like several other members of the cod family, can be seen here too, although mature individuals over 1.5 metres in length typically live well below the limits of sport diving. When not actively hunting other fish, they tend to hover in caves or crevices amongst rocks or on wrecks, where their jutting white chin barbel catches the eye.

Also in or around crevices are crawfish (*Palinurus elephas*), a species of spiny lobster. Distinct from both true lobsters and freshwater crayfish, the crawfish has small front claws and a dark, reddish carapace covered in short spines. Lacking the defence of huge claws, it has another strategy for avoiding

*This page, from top to bottom.*
*The beautiful sunset coral (*Leptopsammia pruvoti*) is one of the few true corals in northern seas; its living tissues obscure a stony skeleton.*

*The large sea anemone* Bolocera tuediae *on a bed of brittle stars (*Ophiothrix fragilis*).*

*Ross coral (*Pentapora foliacea*) is common on deeper sites but also occurs in fairly shallow waters. A coating of minute hydroids gives it a slightly furry appearance.*

*Facing page.*
*A lobe of the soft coral, red sea fingers (*Alcyonium glomeratum*), which occurs on deep rock faces or beneath overhangs, unlike the current-swept sites favoured by its more common relative, dead-man's fingers.*

danger: flexing its muscular abdomen and broad tail fan enables it to shoot backwards when disturbed, usually into a nearby crevice. Crawfish are caught commercially with trammel or tangle nets and pots, and are found from southern Norway to Morocco and the Mediterranean.

Deeper mud slopes, too, have a different fauna to that of the shallows. Salmon-pink Dublin Bay prawns (*Nephrops norvegicus*), commonly trawled from 100 to 300 metres to be marketed as scampi or langoustines, make burrows in fine mud and sit at the entrance with their long front claws projecting. Various other animals typically live in soft sea beds at 20 to 30 metres or more, such as the sea cucumber (*Paracucumaria hyndmani*), and the strangely flattened goosefoot starfish (*Anseropoda placenta*), which is frequently found in sea lochs. Beautiful sea pens, related to soft corals, project from the mud, looking uncannily like large, ornate, sometimes luminescent quills.

## Northern corals

One of the most colourful corals in northern seas, the sunset coral (*Leptopsammia pruvoti*), is strikingly similar to several tropical representatives of its family, which also have a deep yellow column and a mass of translucent tentacles densely beaded in the same colour. Although its appearance is very similar to that of a sea anemone, the outer tissues conceal a stony skeleton and, like the somewhat shallower living Devonshire cup coral, it is a true coral. Known also as the sunset star coral or sunset cup coral, it is primarily Mediterranean, and its occurrence in more northern waters is limited to a few sites in southwestern Britain, notably Lundy Island. It grows up to 5 cm across, as a solitary individual or in small groups under overhangs, in sheltered rock crevices or on deep walls where it will be spared strong water movement.

Equally spectacular are the soft corals called red sea fingers (*Alcyonium glomeratum*), which are found on sheltered rock faces mainly in southwest England and Wales. These are deeper living and much rarer relatives of dead-man's fingers, and have a more exotic appearance with slender, red or orange lobes and contrasting white, star-like polyps.

Like soft corals, the sea fan (*Eunicella verrucosa*) has eight feathery tentacles on each of the small polyps that make up the colony, but it is a gorgonian coral with a harder internal structure. The orange-pink colony grows up to 40 cm across, its network of many branches usually aligned in one plane across the current to aid filter feeding. A tiny sea slug, *Tritonia nilsodhneri*, preys on it, mimicking a branchlet to such perfection that I have found it only after much careful searching. The coiled egg masses – also pink – are laid on the branches too. Growth of the sea fan is slow, perhaps one centimetre per year, and the brittle structure is easily smashed. Accidental damage is not the only threat to contend with: sea fans are plundered by souvenir hunters. *E. verrucosa* occurs on rock from northwest Africa to northwest Ireland, replaced further north by a less conspicuous sea fan with fewer branches, *Swiftia pallida*.

At first sight, ross or rose coral (*Pentapora foliacea*) appears coral-like, but it is actually a bryozoan, related to the sea mat that forms an encrustation on kelp, although with a very different growth form. The large, rounded colonies may grow to half a metre or more across and are composed of rosettes of very brittle, convoluted, leaf-like plates, orange or buff, often with dark green on the less exposed basal part; the whole

*Facing page. The disproportionately tiny body is suspended high above the ground, atop eight stilts, as the sea spider* Nymphon *picks its way in giant strides over an undergrowth of tufts of bryozoans.* Nymphon *species are found at a range of depths in Europe and in the North American Pacific.*

*Overleaf. The large red cushion star (*Porania pulvillus*), here amongst small purple patches of calcareous algae on a steep rock face, grows up to about 11 cm across. A related species called the badge sea star (*Porania insignis*) is found in North American waters from Cape Cod to Hatteras.*

*Below. Deep-living basket stars (*Gorgonocephalus*) are a rare sight in northern European seas. This individual, with its branching arms partly extended, spans about 20 cm. In North America, they are not confined to deep reefs, and species such as the northern basket star (*G. arcticus*) – in the Atlantic from the Arctic to Cape Cod – and* G. eucnemis *– in the Pacific from the Bering Sea to California – are regularly seen by divers.*

structure is reminiscent of an ornamental cabbage. It is only on very close examination of the plates that the relationship to the encrusting sea mat becomes apparent. The typical bryozoan network of compartments, given a slightly fluffy coating by the fine tentacles of the individual animals, can be seen with a magnifying lens. Small hydroids commonly grow on ross coral, particularly near the outer rim of the plates, while other tiny animals, including cowries and the hairy or bristly crab (*Pilumnus hirtellus*), take shelter between the plates. Ross coral is found attached to rock or sometimes stones from the Atlantic coasts of southwest Britain to the Mediterranean.

## *Deep stars*

The name cushion star is given to several species of starfish that have a large body and five short tapered arms, but to none so aptly as to the large red cushion star, *Porania pulvillus*. The colour of the plump fleshy body can be startling, rich scarlet enhanced by the velvety appearance of the upper surface (though this is said to be greasy to touch), and by the tapestry-like effect of the clusters of small, soft, whitish papulae or gills radiating from the centre and interspersed with fine lines or furrows. The red cushion star is usually found at depths of about 20 metres or more on current-swept rocks (where it is thought to include dead-man's fingers in its diet), as well as at more sheltered sites. Although known as far south as the Bay of Biscay, it is more common further north, from northern Britain to Norway.

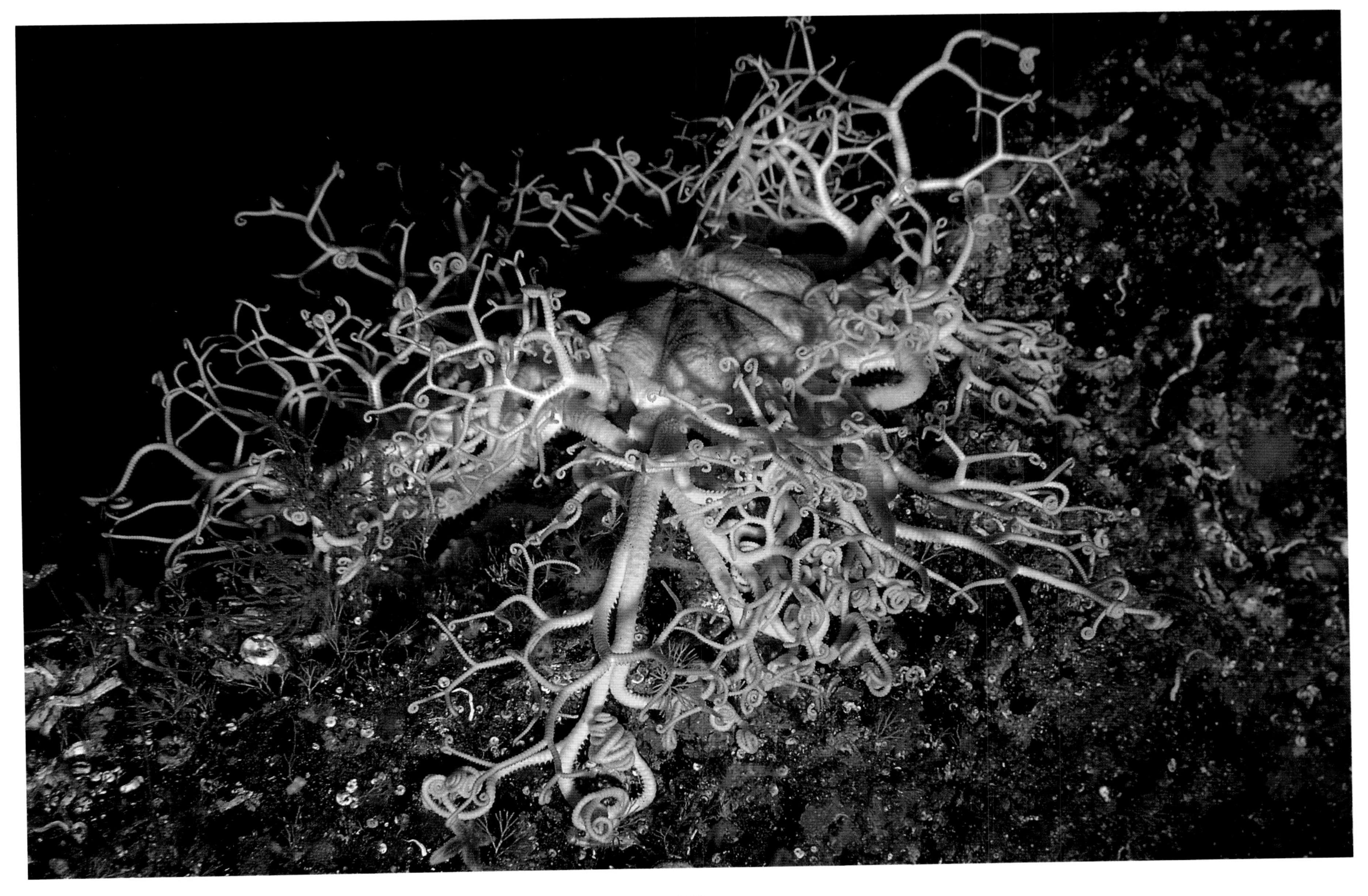

Deeper sites may offer the chance to see some highly unusual and spectacular relatives of starfish: basket stars. One dive down a sheer rock face in a Norwegian fjord was planned to 35 metres, deeper than my normal preference, but for once I was happy to settle for minutes rather than an hour or more. Just below a narrow ledge we halted rather than continuing down the seemingly endless wall. Here was what we had come for, first one, then several more basket stars, each a tangle of whitish and buff-coloured branched arms. The arms were unfurled, apart from the tips coiled like the tendrils of a fern. Several species of basket star (*Gorgonocephalus*) live on rock or other ground off northern Europe but although they may occur in large aggregations they are typically deep-water species rarely seen by divers. Their branched arms trap tiny plankton from the current and then pass the food to the central mouth. Basket stars are related to brittle stars, which are common on shallower reefs.

Until a year or so before our visit there had been colonies of a rare soft coral at the same place on the wall, but they had now died out except for possible survivors at greater depths; local opinion was that they might have been affected by chemicals used at a nearby salmon farm.

One other find did await us. Glancing around the area hurriedly, conscious of our limited time, we saw a very different animal, the large sea spider, *Nymphon gracile*. Many sea spiders (not true spiders but members of another arthropod group, the Pycnogonida) are small and inconspicuous, but these animals were fist-sized, including their exaggeratedly spindly legs. Reluctantly we acknowledged that it was time to leave and began our slow ascent up the wall.

# CHAPTER NINE

# THE OPEN SEA

*Below. Adult horse mackerel (*Trachurus trachurus*) are fast-swimming predators of smaller shoaling fish in open water. Their juveniles, sometimes seen around the tentacles of jellyfish, prey instead on minute planktonic invertebrates. Horse mackerel are related to jack mackerel (*T. symmetricus*), a North American species in the Pacific from British Columbia to Baja California.*

In open water a variety of animals – both large and small – are carried with the ocean currents, drifting wherever these take them. Near the surface, jellyfish can be seen pulsating in a slow rhythm. Although they may appear to be entirely at the sea's mercy, they are able to propel themselves slowly, usually upwards, by contracting and dilating the bell. Among the many planktonic animals that live suspended in the water are comb jellies, copepods (minute crustaceans) and crustaceans and other invertebrates in their larval stages.

Plankton provide a rich source of food for various animals whose lifestyles are adapted to the open sea. Atlantic herring (*Clupea harengus*) live in open water but are likely to be seen near the surface only at night. They and other members of their family, small silvery shoaling fishes such as sprats and pilchards, feed mainly on planktonic crustaceans, and both crustaceans and fish are preyed on by fast-swimming squids. The herring family are eaten by elongated garfish (*Belone belone*), which catch their small prey sideways in their long beaks. Atlantic mackerel (*Scomber scombrus*) also feed on young herrings and sprats, as well as on plankton, and sometimes swim in enormous schools, while juvenile mackerel often shoal with horse mackerel (*Trachurus trachurus*). Other schooling fish, grey mullet (*Chelon labrosus*), are sometimes seen inshore but

*The basking shark* (Cetorhinus maximus) *skims the surface with its dorsal fin, keeping its mouth wide open to take in plankton. Its wide distribution in temperate seas includes most of Europe and the North American Pacific.*

*Photographs by Brian Pitkin*

spawn in open water. Some animals that cruise the surface are considerably larger, and a sighting can be a real thrill, whether it is of a basking shark (a cooler water counterpart of the tropical whale shark), a sunfish, dolphins or seals.

Encounters are often unexpected. As our inflatable boat heads towards a dive site, one of our group sometimes sights an object in the water, and we make a detour to inspect it. It might turn out to be mere flotsam or jetsam, but occasionally it is something more exciting. In Cornwall recently, we were fortunate to see a leatherback (*Dermochelys coriacea*), the largest of the turtles and a predator of jellyfish. This is a threatened species worldwide and seldom reaches European coasts.

Many animals are wary, and will submerge and swim away even if a snorkeller takes care to slip into the water making the least splash possible. Jellyfish, though, are not disturbed by the presence of humans in the water and can be observed and photographed at leisure, or rather for as long as one's fellow divers are prepared to wait. Sometimes you get an extra opportunity while surfacing at the end of a dive.

## *Jellyfish*

The shapeless lumps of jelly that can look so unattractive washed up on the beach have none of the delicate beauty of living jellyfish floating in the sea; the cosmopolitan moon jellyfish (*Aurelia aurita*), for example, is a mesmerising creature. Its near-transparent, bluish-white-tinged bell, an inverted saucer up to

*Right. Juvenile fish sheltering amongst the tentacles of the lion's-mane jellyfish* (Cyanea)*. Two* Cyanea *species occur in the European Atlantic and the larger of these,* C. capillata, *is also found along both coasts of North America.*

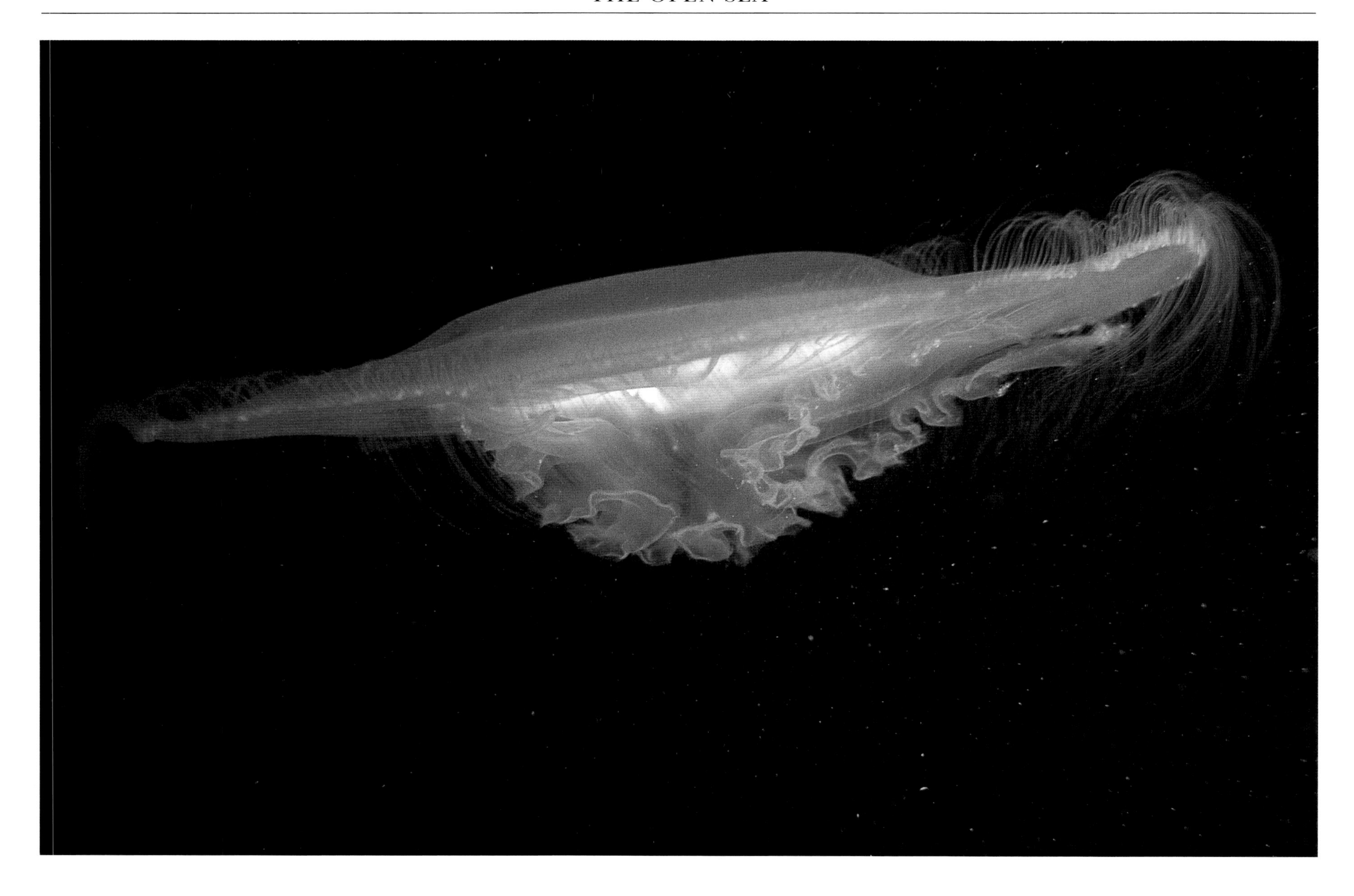

about 30 cm in diameter, is veined with faint, purple lines that form part of the animal's digestive system. The hair-like veil of numerous fine, short tentacles fringing the bell is not as innocuous as it appears; although not a danger to swimmers, it bears the stinging cells that stun the jellyfish's small prey. Additional food is trapped on the mucus covering the bell. Frilly mouth arms hang down below, and beneath the centre of the bell are four, distinctive, pale violet circles or crescents, the gonads, clearly visible through the bell from above.

The compass jellyfish (*Chrysaora hysoscella*) has a similarly sized bell but with a fringe of longer, stouter tentacles, and longer mouth arms. It takes its common name from the distinctive, reddish-brown lines radiating from the centre of the bell. It can deliver a sting, but not so severe as that of a circumpolar species, the lion's-mane (*Cyanea capillata*), a large brownish-purple jellyfish with a mass of tentacles hanging from its bell. These tentacles may extend for two metres, trawling for small organisms. I have been stung painfully on the lip when I thought I was clear of the jellyfish, but had failed to notice one of its trailing tentacles brush my face, catching me between mask and diving regulator. A related species, *Cyanea lamarckii*, is smaller and more bluish.

The barrel jellyfish (*Rhizostoma pulmo*) is huge and surprisingly solid, its near-opaque, milky dome up to 90 cm in diameter. Despite this impressive appearance, the outer parts are harmless to touch: unlike many jellyfish, they have no tentacles armed with stinging cells. Instead, cauliflower-shaped structures on the

*Differing aspects of the moon jellyfish (*Aurelia aurita*). The image above shows the frilly mouth arms below the bell; in the picture opposite, the gonads can be seen through the bell. The widespread range of the moon jellyfish includes the European Atlantic, across to Greenland, the Arctic, and Alaska, and down both coasts of North America, to the West Indies and California.*

mouth arms sieve out the jellyfish's nourishment of tiny plankton from the water. The eight mouth arms are joined for part of their length and hang stiffly beneath the lobed purple rim of the bell. Like several other jellyfish, it is frequently accompanied by juvenile horse mackerel, which probably benefit from its protection. Typically a southern species, the barrel jellyfish is at its most common in the Mediterranean and off English coasts, but is known as far north as Norway.

The string jelly (*Apolemia uvaria*) is not strictly a jellyfish but a siphonophore, a group that also includes the infamous Portuguese man-of-war. Siphonophores are colonies of individuals which in many species are kept afloat by a bladder. In the string jelly this float is small and at one end of the trailing stem that links the transparent individuals. This strange-looking colony may be immensely long – up to 20 metres.

## *Plankton feeders*

A steel-grey, triangular shark's fin cut the surface as it headed towards my husband who was snorkelling. He watched, bobbing in the water with mounting excitement as it drew near. The shark's bulk and length was impressive, but at 5 metres nowhere near the record of 11 metres or more. At the last moment it veered aside, allowing Brian to photograph its flanks as it passed. The huge mouth gaped constantly but revealed no saw-toothed menace, for this was a harmless giant, the plankton-feeding basking shark (*Cetorhinus maximus*). Even so, its sheer size demands a little caution, as Brian discovered when a retreating tail fin struck him a glancing blow.

As the shark swims, its open mouth, beneath a curiously shaped, upturned snout, takes in water. Gill rakers strain out vast numbers of tiny crustaceans before the water is expelled through the gill slits, five long slashes on each side. The diet on which the shark thrives could prove to be its downfall. Squalene, stored as a dietary by-product in its large, oil-rich liver, is sought after for many uses, from cosmetics to aircraft lubricant, and there are fears that commercial fishing may already be putting populations at risk. There is an urgent need for more information about the animal's life history, before it will be possible to recommend sustainable levels for kills.

Brian had made a special trip to the Isle of Man at short notice as soon as he had heard of the arrival of the basking sharks, which are regular summer visitors to this and other northeastern Atlantic coasts. Various animals migrate inshore from oceanic waters. On one occasion in Wales when a fin was spotted from our dive boat, Brian prepared to jump into the sea. He changed his mind at the last moment, though, when the shark was identified as a blue shark, smaller but potentially dangerous – at least in open, tropical waters. Although primarily a warm-water species, the pelagic blue shark (*Prionace glauca*) is sometimes seen swimming slowly at the surface as far north as Britain, where it is a prized target for sport angling; it has even, occasionally, been sighted off Norwegian coasts. By night it is a fast-swimming hunter of squid and shoaling fish.

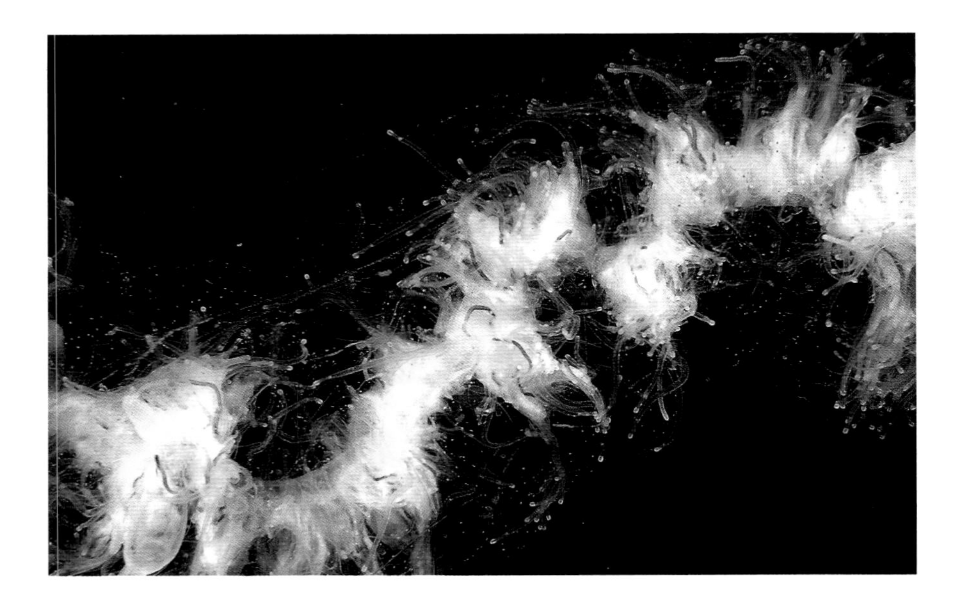

*Left. A small section of the long, string jelly* (Apolemia uvaria) *shows the densely grouped polyps that make up this unusual siphonophore.*

*Facing page. Beneath the bell of the barrel jellyfish* (Rhizostoma pulmo), *its sturdy mouth arms take in plankton through many orifices.*

The dorsal fin of another large plankton feeder sticks out of the water as it rests, drifting at the surface. The ocean sunfish (*Mola mola*) is an uncommon sight around northern European coasts but it is widespread in warm and temperate seas. It can grow to an impressive three metres and has a strangely truncated appearance, with a body that ends abruptly at the long dorsal and anal fins, and a tail reduced to a short frill.

## *Marine mammals*

For most people in northern Europe, whales are a rare sight. Those most likely to be seen are the engaging, intelligent and unafraid bottle-nosed dolphin (*Tursiops truncatus*) and the

*Below. Long, frilly mouth arms trail beneath the compass jellyfish* (Chrysaora hysoscella) *as it drifts near the surface in the coastal waters of Atlantic Europe. It is related to the sea nettle* (C. quinquecirrha), *in the North American Atlantic from Cape Cod to the West Indies, and the brown jellyfish* (C. melanaster), *in the Pacific from Alaska to California.*

common or harbour porpoise (*Phocoena phocoena*), which is gregarious with its own kind but wary of people. Diving with seals is much easier to arrange and an experience not to be missed. The plump, grey-brown shapes on the rocks above the water stir at the approach of our small boat and raise dog-like heads to inspect us. Some continue to laze, while others slip into the water to join those already there, heads bobbing at the surface. As we dive, we glimpse the shapes again, now transformed with speed and grace as they twist and turn through weeds and rocky gullies. Grey seals (*Halichoerus grypus*) are a delight to see underwater, inquisitive and playful. The games are on their own terms, though. They seem to be attracted to divers' fins, at least if these are colourful, and will often give them a trial nip or a gentle tug. The seal's usual approach is from behind; as I swing round, my camera at the ready, the animal rolls and dives away. Sometimes a backwards somersault or two will entice the seal back to the diver's fins again.

There are populations of grey seals on both sides of the Atlantic and in the Baltic. In northern Europe they are especially numerous in British waters, where I have seen them at sites including St Kilda, the Farne Islands and the Isle of Man. The western Atlantic grey seals live principally from Labrador to Nantucket, and sometimes further south. More abundant than the common or Atlantic harbour seal (*Phoca vitulina*) in Europe, they were less badly affected when a viral epidemic struck seals in the North Sea in 1988. Seals use their agility in the water to catch the vast quantities of fish they consume, but this puts them in direct competition with the fishing industry. Such conflicts underline the need to find ethical and practical solutions to the problems caused by our impact, if we are to safeguard the sea's resources.

*A grey seal (*Halichoerus grypus*), swimming over seaweed-covered gullies, pauses to investigate me as I move in to photograph it. The Canadians sometimes call grey seals horseheads.*

# PHOTOGRAPHIC DETAILS

Details of the equipment and settings I used for the photography in this book are given below. Unless stated otherwise, the film stock was Fujichrome Velvia, and a Sunpak Marine 32 flashgun (strobe) with a diffuser was used. Nearly all of the photographs were taken at 1/60 second.

*Endpapers.* England: SE. Cornwall, Whitsand Bay
Nikonos V with 28 mm lens plus close-up lens; f16-22; Kodachrome 64

*Half title.* England: Dorset, Kimmeridge
Nikonos V with 35 mm lens; extension tube 1:2; f16-22

*Title spread.* England: Devon, Lundy Island
Nikonos III with 15 mm lens; f8; Ektachrome 100

*Page 4.* Norway: SW. of Kristiansund, Strømsholmen
Nikonos V with 35 mm lens; extension tube 1:2; f16-22

*Page 5.* Scotland: Berwickshire, St Abbs
Nikonos III with 28 mm lens plus close-up lens; f16-22; Kodachrome 64

*Page 6, left.* England: SE. Cornwall, Penlee Point
Nikonos III with 15 mm lens; f8-11; Agfachrome 100

*Page 6, right.* England: Sussex, Shoreham
Pentax LX with 50 mm macro lens; Hugyfot housing; f11-16

*Page 7.* Norway: SW. of Kristiansund, Strømsholmen
Nikonos V with 15 mm lens; f8; Fujichrome Sensia 100

*Page 8, top.* England: Sussex, Pagham
Nikonos V with 28 mm lens; f8-11; Kodachrome 64

*Page 8, bottom.* England: Devon, Plymouth Sound
Nikonos V with 35 mm lens; extension tube 1:2; f16

*Page 9.* Scotland: St Kilda
Nikonos V with 28 mm lens plus close-up lens; f16-22; Kodachrome 64

*Page 10.* England: Devon, Lundy Island
Nikonos III with 15 mm lens; f8-11; Agfachrome 100

*Page 11, left.* England: Dorset, Kimmeridge
Nikonos V with 35 mm lens; extension tube 1:2; f16-22

*Page 11, right.* Wales: Dyfed, Skokholm Island
Nikonos III with 35 mm lens; extension tube 1:2; f16-22; Kodachrome 64

*Page 12, top.* England: Devon, Plymouth Sound
Nikonos V with 35 mm lens; extension tube 1:2; f16-22

*Page 12, bottom.* Isle of Man: Gantry Ledges
Pentax LX with 50 mm macro lens; Hugyfot housing; f11-16

*Page 13, top.* England: Sussex, Bognor
Pentax LX with 50 mm macro lens; Hugyfot housing; f16

*Page 13, bottom.* England: Devon, Plymouth Sound
Nikonos V with 28 mm lens plus close-up lens; f16-22

*Page 14.* Norway: SW. of Kristiansund, Strømsholmen
Nikonos V with 35 mm lens; extension tube 1:2; f16-22

*Page 15.* England: Dorset, Swanage
Nikonos V with 35 mm lens; extension tube 1:2; f16-22

*Page 16, top to bottom.*

Norway: SW. of Kristiansund, Strømsholmen
Nikonos V with 35 mm lens plus close-up lens; f11-16

Norway: SW. of Kristiansund, Strømsholmen
Pentax LX with 50 mm macro lens; Hugyfot housing; f11-16

England: Devon, Plymouth Sound
Nikonos V with 35 mm lens; extension tube 1:2; f16

Isle of Man: Sugarloaf Caves
Nikonos V with 35 mm lens; extension tube 1:2; f16

*Page 17, top.* England: Devon, Plymouth Sound
Nikonos V with 35 mm lens; extension tube 1:2; f16-22; Kodachrome 64

*Page 17, bottom.* Norway: SW. of Kristiansund, Strømsholmen
Nikonos V with 35 mm lens; extension tube 1:1; f16

*Pages 18-19.* England: Devon, Lundy Island
Nikonos V with 28 mm lens plus close-up lens; f16-22; Kodachrome 64

*Pages 20-21.* England: Sussex, Shoreham
Nikonos II with 35 mm lens; extension tube 1:2; Vivitar flash; f22; Kodachrome 64

*Page 21, bottom.* England: Sussex, Pagham
Pentax LX with 50 mm macro lens; Hugyfot housing; f11-16

*Pages 22-23.* England: Sussex, Pagham
Pentax LX with 50 mm macro lens; Hugyfot housing; f16

*Page 24.* Norway: SW. of Kristiansund, Strømsholmen
Pentax LX with 50 mm macro lens; Hugyfot housing; f16

*Page 25.* England: Cornwall, Lamorna Cove
Nikonos V with 28 mm lens plus close-up lens; f16-22

*Pages 26-27.* England: Sussex, Shoreham
Pentax LX with 50 mm macro lens; Hugyfot housing; f16

*Page 28, top and centre.* England: Cornwall, Lamorna Cove
Nikonos V with 28 mm lens plus close-up lens; f22

*Page 28, bottom.* England: Cornwall, Logan Rock
Pentax LX with 50 mm macro lens; Hugyfot housing; f11-16

*Page 29.* Norway: SW. of Kristiansund, Strømsholmen
Nikonos V with 35 mm lens; extension tube 1:2; f16-22

*Page 30.* Norway: SW. of Kristiansund, Strømsholmen
Nikonos V with 28 mm lens; f8

*Pages 30-31.* Scotland: Berwickshire, St Abbs
Nikonos V with 28 mm lens plus close-up lens; f22; Ektachrome 64

*Pages 32-33.* Norway: SW. of Kristiansund, Strømsholmen
Pentax LX with 50 mm macro lens; Hugyfot housing; f11

*Page 34, top to bottom.*

England: Sussex, Pagham
Pentax LX with 50 mm macro lens; Hugyfot housing; f16

England: Cornwall, Lamorna Cove
Nikonos V with 28 mm lens plus close-up lens; f16-22

England: Cornwall, Lamorna Cove
Nikonos V with 28 mm lens plus close-up lens; f16-22

England: Sussex, Bognor
Pentax LX with 50 mm macro lens; Hugyfot housing; f16

*Page 35.* England: Devon, Salcombe
Pentax LX with 50 mm macro lens; Hugyfot housing; f8-11

*Page 36 and Page 37.* Norway: SW. of Kristiansund, Strømsholmen
Nikonos V with 35 mm lens; extension tube 1:2; f16-22

*Pages 38-39.* Norway: SW. of Kristiansund, Strømsholmen
Pentax LX with 50 mm macro lens; Hugyfot housing; f11-16

*Page 40.* Norway: SW. of Kristiansund, Strømsholmen
Nikonos V with 35 mm lens; extension tube 1:2; f16-22

*Page 41, top left.* Norway: SW. of Kristiansund, Strømsholmen
Pentax LX with 50 mm macro lens; Hugyfot housing; f11-16

*Page 41, bottom left.* England: Cornwall, The Manacles
Pentax LX with 50 mm macro lens; Hugyfot housing; f11-16

*Page 41, top right.* Scotland: Sound of Mull, Dearg Sgeir
Nikonos III with 28 mm lens plus close-up lens; f16-22; Kodachrome 64

*Page 41, bottom right.* Wales: Dyfed, Skomer Island
Nikonos with close-up lens; f16-22; Kodachrome 64

*Page 42.* England: Devon, Lundy Island
Nikonos II with 35 mm lens; extension tube 1:2; Vivitar flash; f22; Kodachrome 64

*Page 43.* Norway: SW. of Kristiansund, Strømsholmen
Nikonos V with 15 mm lens; f8; Fujichrome Sensia 100

*Page 44.* England: Dorset, Swanage
Nikonos V with 35 mm lens; extension tube 1:1; f16

*Page 45, top.* England: Cornwall, Longships Lighthouse
Pentax LX with 50 mm macro lens; Hugyfot housing; f11-16

*Page 45, bottom.* England: Dorset, Kimmeridge
Nikonos V with 35 mm lens; extension tube 1:2; f16-22

*Page 46.* Scotland: Berwickshire, St Abbs
Nikonos III with 28 mm lens plus close-up lens; f16-22; Kodachrome 64

*Page 47.*England: Devon, Salcombe
Pentax LX with 50 mm macro lens; Hugyfot housing; f11-16

*Page 48, top.* England: Devon, Plymouth Sound
Nikonos V with 35 mm lens; extension tube 1:2; f16-22

*Page 48, bottom.* England: Devon, Lundy Island
Nikonos III with 35 mm lens; extension tube 1:1; f16-22; Kodachrome 64

*Page 49, bottom.* Wales: Dyfed, Skomer Island
Nikonos II with 35 mm lens; extension tube 1:2; Vivitar flash; f16-22; Kodachrome 64

*Page 50, left.* Scotland: Berwickshire, St Abbs
Nikonos III with 28 mm lens; f8; Kodachrome 64

*Pages 50-51.* England: Devon, Plymouth Sound
Nikonos V with 35 mm lens; extension tube 1:2; f16-22; Kodachrome 64

*Page 51, right.* Wales: Dyfed, E. of Chapel Rocks
Nikonos III with 28 mm lens plus close-up lens; f22; Ektachrome 100

*Page 52, top.* England: Cornwall, Logan's Gully
Pentax LX with 50 mm macro lens; Hugyfot housing; f16

*Page 52, middle.* England: Sussex, Littlehampton
Nikonos III with 35 mm lens plus close-up lens; f16-22; Kodachrome 64

*Page 52, bottom left.* England: Devon, Plymouth Sound
Nikonos V with 28 mm lens plus close-up lens; f16-22

*Page 52, bottom right.* England: Sussex, Bognor
Nikonos V with 35 mm lens; extension tube 1:2; f16-22

*Page 53, top left.* England: Devon, Plymouth Sound
Nikonos V with 35 mm lens; extension tube 1:2; f16-22

*Page 53, top right.* Isle of Man: St Michael's Isle (Fort Island)
Pentax LX with 50 mm macro lens; Hugyfot housing; f16

*Page 53, bottom.* Scotland: Berwickshire, St Abbs
Nikonos III with 28 mm lens; f11; Kodachrome 64

*Page 54, top.* England: Cornwall, Logan's Gully
Pentax LX with 50 mm macro lens; Hugyfot housing; f11-16

*Page 54, bottom.* Scotland: Berwickshire, St Abbs
Nikonos III with 28 mm lens plus close-up lens; f16-22; Kodachrome 64

*Page 55, centre right.* England: Cornwall, The Manacles
Pentax LX with 50 mm macro lens; Hugyfot housing; f11-16

*Page 55, bottom left.* Norway: SW. of Kristiansund, Strømsholmen
Pentax LX with 50 mm macro lens; Hugyfot housing; f11

*Page 55, bottom right.* Norway: SW. of Kristiansund, Strømsholmen
Pentax LX with 50 mm macro lens; Hugyfot housing; f11-16

*Pages 56-57.* Scotland: Berwickshire, St Abbs
Nikonos III with 28 mm lens plus close-up lens; f16-22; Kodachrome 64

*Page 57, right.* Norway: SW. of Kristiansund, Strømsholmen
Nikonos V with 35 mm lens; extension tube 1:2; f16-22

*Page 58, top.* Norway: SW. of Kristiansund, Strømsholmen
Pentax LX with 50 mm macro lens; Hugyfot housing; f16

*Page 58, centre.* England: Devon, Plymouth Sound
Pentax LX with 50 mm macro lens; Hugyfot housing; f8-11

*Page 58, bottom.* Norway: SW. of Kristiansund, Strømsholmen
Pentax LX with 50 mm macro lens; Hugyfot housing; f8-11

*Page 59.* Norway: SW. of Kristiansund, Strømsholmen
Pentax LX with 50 mm macro lens; Hugyfot housing; f11-16

*Page 60, top.* Wales: Dyfed, Milford Haven
Nikonos III with 35 mm lens; extension tube 1:2; f16-22; Kodachrome 64

*Page 60, bottom.* Isle of Man: St Michael's Isle (Fort Island)
Pentax LX with 50 mm macro lens; Hugyfot housing; f16

*Page 61, top.* England: Devon, Plymouth Sound
Nikonos V with 35 mm lens; extension tube 1:2; f16-22; Ektachrome Professional 64

*Page 61, bottom left.* Wales: Dyfed, Gateholm Bay
Nikonos III with 35 mm lens; extension tube 1:2; f16-22; Kodachrome 64

*Page 61, bottom right.* Wales: Dyfed, West Blockhouse Point
Nikonos II with 35 mm lens; extension tube 1:2; Vivitar flash; f22; Ektachrome 64

*Page 62, bottom left.* Scotland: St Kilda
Nikonos III with 35 mm lens; extension tube 1:2; f16-22; Kodachrome 64

*Pages 62-63.* Norway: SW. of Kristiansund, Strømsholmen
Nikonos V with 35 mm lens; extension tube 1:2; f16-22

*Page 64.* Norway: SW. of Kristiansund, Strømsholmen
Pentax LX with 50 mm macro lens; Hugyfot housing; f11-16

*Page 65, top.* Scotland: St Kilda
Nikonos V with 35 mm lens; extension tube 1:2; f16-22; Kodachrome 64

*Page 65, bottom.* Norway: SW. of Kristiansund, Strømsholmen. Pentax LX with 50 mm macro lens; Hugyfot housing; f11-16

*Pages 66-67.* England: Devon, Salcombe. Nikonos III with 15 mm lens; f8; Kodachrome 64

*Page 67, right.* England: Devon, Plymouth Sound
Pentax LX with 50 mm macro lens; Hugyfot housing; f11-16

*Page 68.* England: SE. Cornwall, Whitsand Bay
Nikonos V with 28 mm lens plus close-up lens; f16-22

*Page 69.* Norway: SW. of Kristiansund, Strømsholmen
Pentax LX with 50 mm macro lens; Hugyfot housing; f11

*Page 70.* Norway: SW. of Kristiansund, Strømsholmen
Nikonos V with 35 mm lens; extension tube 1:2; f16-22

*Page 71, top.* Norway: SW. of Kristiansund, Strømsholmen
Nikonos V with 35 mm lens; extension tube 1:2; f16-22; or Pentax LX with 50 mm macro lens; Hugyfot housing; f16

*Page 71, bottom.* Norway: SW. of Kristiansund, Strømsholmen
Nikonos V with 35 mm lens; extension tube 1:1; f16

*Pages 72-73.* England: Sussex, Pagham
Nikonos V with 28 mm lens; f8-11

*Page 73, top right.* England: Sussex, Pagham
Nikonos II with 28 mm lens; Vivitar flash; f11; Ektachrome 64

*Page 74.* England: Dorset, Swanage
Nikonos V with 35 mm lens; extension tube 1:2; f16-22

*Page 75.* Isle of Man: Sugarloaf Caves
Nikonos V with 35 mm lens; extension tube 1:2; f16-22

*Pages 76-77.* Isle of Man: Sugarloaf Caves
Nikonos V with 35 mm lens; extension tube 1:2; f16

*Page 76, bottom.* Scotland: Sound of Mull, Calve Island
Nikonos V with 35 mm lens; extension tube 1:2; f16-22; Kodachrome 64

*Page 77, bottom.* Scotland: Sound of Mull, Dearg Sgeir
Nikonos III with 28 mm lens plus close-up lens; f16-22; Kodachrome 64

*Pages 78-79.* Scotland: St Kilda
Nikonos V with 28 mm lens plus close-up lens; f16-22; Kodachrome 64

*Page 80.* England: Cornwall, Runnel Stone
Nikonos V with 35 mm lens; extension tube 1:1; f16

*Page 81, top left.* England: SE. Cornwall, Whitsand Bay
Nikonos V with 28 mm lens plus close-up lens; f16

*Page 81, top right.* England: Devon, Lundy Island
Nikonos V with 28 mm lens plus close-up lens; f16-22; Kodachrome 64

*Page 81, bottom.* Norway: SW. of Kristiansund, Strømsholmen
Nikonos V with 35 mm lens plus close-up lens; f16

*Page 82, top.* England: Sussex, Bognor
Nikonos V with 35 mm lens; extension tube 1:1; f16

*Page 82, bottom.* England: SE. Cornwall, Whitsand Bay
Nikonos V with 28 mm lens; f8-11; Kodachrome 64

*Page 83.* Norway: SW. of Kristiansund, Strømsholmen
Nikonos V with 15 mm lens; f8; natural light; Fujichrome Sensia 100

*Pages 84-85.* Norway: SW. of Kristiansund, Strømsholmen
Nikonos V with 15 mm lens; f8; Fujichrome Sensia 100

*Page 86, top.* Norway: SW. of Kristiansund, Strømsholmen
Nikonos V with 35 mm lens; extension tube 1:2; f16-22

*Page 86, centre.* Norway: SW. of Kristiansund, Strømsholmen
Pentax LX with 50 mm macro lens; Hugyfot housing; f16

*Page 86, bottom left.* Norway: SW. of Kristiansund, Strømsholmen
Nikonos V with 35 mm lens; extension tube 1:1; f16

*Page 86, bottom right.* Norway: SW. of Kristiansund, Strømsholmen
Nikonos V with 35 mm lens; extension tube 1:2; f16-22

*Page 87.* Norway: SW. of Kristiansund, Strømsholmen
Nikonos V with 35 mm lens; extension tube 1:2; f16-22

*Page 88, left.* Norway: SW. of Kristiansund, Strømsholmen
Nikonos V with 35 mm lens; extension tube 1:2; f16

*Pages 88-89.* Scotland: Sound of Mull, Calve Island
Nikonos V with 35 mm lens; extension tube 1:2; f16-22; Kodachrome 64

*Page 90, top.* England: Devon, Plymouth Sound
Nikonos V with 35 mm lens; extension tube 1:2; f16-22

*Page 90, bottom.* Norway: SW. of Kristiansund, Strømsholmen
Pentax LX with 50 mm macro lens; Hugyfot housing; f16

*Page 91.* England: Devon, Plymouth Sound
Pentax LX with 50 mm macro lens; Hugyfot housing; f11-16

*Pages 92-93.* Isle of Man: St Michael's Isle (Fort Island)
Pentax LX with 50 mm macro lens; Hugyfot housing; f11

*Page 93, bottom.* England: Sussex, Selsey
Pentax LX with 50 mm macro lens; Hugyfot housing; f16

*Page 94, bottom.* Isle of Man: Port Erin
Pentax LX with 50 mm macro lens; Hugyfot housing; f11

*Page 95.* Norway: SW. of Kristiansund, Strømsholmen
Nikonos V with 15 mm lens; f8; Fujichrome Sensia 100

*Pages 96-97.* Isle of Man: Port Erin
Pentax LX with 50 mm macro lens; Hugyfot housing; f11; Agfachrome 50

*Page 97, bottom.* Norway: SW. of Kristiansund, Strømsholmen
Nikonos V with 35 mm lens; extension tube 1:2; f16-22

*Page 98.* England: Dorset, Kimmeridge
Nikonos V with 35 mm lens; extension tube 1:2; f16-22

*Page 99.* England: Devon, Plymouth Sound
Nikonos V with 35 mm lens; extension tube 1:2; f16-22; Kodachrome 64

*Page 100, top.* England: Dorset, Swanage
Nikonos V with 35 mm lens; extension tube 1:2; f16-22

*Page 100, bottom.* England: Sussex, Pagham
Pentax LX with 50 mm macro lens; Hugyfot housing

*Page 101.* Norway: SW. of Kristiansund, Strømsholmen
Nikonos V with 35 mm lens plus close-up lens; f16

*Page 102.* England: Sussex, Selsey
Pentax LX with 50 mm macro lens; Hugyfot housing; f11-16

*Page 103.* Scotland: St Kilda
Nikonos V with 28 mm lens plus close-up lens; f16-22; Kodachrome 64

*Page 104.* England: Dorset, Swanage
Nikonos V with 35 mm lens; extension tube 1:2; f16-22

*Page 105, right, top to bottom.*

England: Sussex, Selsey
Pentax LX with 50 mm macro lens; Hugyfot housing; f11-16

England: Sussex, Bognor
Pentax LX with 50 mm macro lens; Hugyfot housing; f11-16

England: Sussex, Bognor
Pentax LX with 50 mm macro lens; Hugyfot housing; f16

England: Sussex, Pagham
Nikonos V with 35 mm lens; extension tube 1:2; f16-22

*Page 105, bottom left.* England: Dorset, Kimmeridge
Nikonos V with 35 mm lens; extension tube 1:2; f16-22

*Page 106, top.* Scotland: Berwickshire, Eyemouth
Nikonos II with 35 mm lens plus Vizmaster lens; Vivitar flash; f16; Ektachrome 200

*Page 106, bottom.* Scotland: St Kilda
Nikonos V with 28 mm lens plus close-up lens; f16-22; Kodachrome 64

*Page 107.* Wales: Dyfed, Milford Haven
Nikonos V with 28 mm lens; f8-11; Kodachrome 64

*Pages 108-109.* England: Devon, Lundy Island
Nikonos V with 35 mm lens; extension tube 1:2; f16-22; Kodachrome 64

*Pages 110-111.* Scotland: Loch Sunart, Sligneach Mor
Nikonos III with 35 mm lens; extension tube 1:2; f16-22; Kodachrome 64

*Page 112.* England: Devon, Plymouth Sound
Nikonos V with 35 mm lens; extension tube 1:2; f16-22

*Page 113, top.* England: Devon, Lundy Island
Nikonos V with 35 mm lens; extension tube 1:2; f16-22; Kodachrome 64

*Page 113, middle.* Scotland: Berwickshire, St Abbs
Nikonos III with 28 mm lens plus close-up lens; f16-22; Kodachrome 64

*Page 113, bottom.* England: Cornwall, near Lamorna Cove
Pentax LX with 50 mm macro lens; Hugyfot housing; f11-16

*Page 114.* Norway: SW. of Kristiansund, Strømsholmen
Pentax LX with 50 mm macro lens; Hugyfot housing; f16

*Page 115.* Norway: SW. of Kristiansund, Strømsholmen
Pentax LX with 50 mm macro lens; Hugyfot housing; f11-16

*Page 116.* Norway: SW. of Kristiansund, Strømsholmen
Pentax LX with 50 mm macro lens; Hugyfot housing; f11-16

*Page 117.* England: SE. Cornwall, Whitsand Bay
Nikonos V with 28 mm lens; f8-11; Kodachrome 64

*Pages 118-119.* Isle of Man
Nikonos V with 15 mm lens; natural light; Kodachrome 64

*Page 118, bottom.* Isle of Man
Nikonos V with 15 mm lens; natural light; Kodachrome 64

*Page 119, bottom right.* Scotland: Berwickshire, St Abbs
Nikonos III with 28 mm lens plus close-up lens; f16; Kodachrome 64

*Page 120 and Page 121.* Isle of Man: near Bows Rocks
Pentax LX with 50 mm macro lens; Hugyfot housing; f8-11

*Page 122, top.* Scotland: St Kilda
Nikonos III with 35 mm lens; extension tube 1:2; f16-22; Kodachrome 64

*Page 122, bottom.* England: Devon, Lundy Island
Nikonos III with 15 mm lens; f11-16; Ektachrome 100

*Page 123.* England: Devon, Plymouth Sound
Nikonos V with 28 mm lens; f8; Agfachrome 50

*Pages 124-125.* Isle of Man: Calf of Man
Nikonos V with 15 mm lens; f5.6-8; Agfachrome 50

*Page 127.* Norway: SW. of Kristiansund, Strømsholmen
Pentax LX with 50 mm macro lens; Hugyfot housing; f16

# BIBLIOGRAPHY

The following list is a small selection of some of the many books that I have found particularly useful or recommend for further reading.


Acker, R.G., Moss, D., & Picton, B.E. *Sponges of the British Isles, A Colour Guide and Working Document* Marine Conservation Society, Ross on Wye, 1992.

Campbell, A. *Hamlyn Guide: Seashores and Shallow Seas of Britain and Europe* Hamlyn, London, 1994

Dipper, F. *British Sea Fishes* Underwater World Publications, London, 1987.

Dipper, F. & Powell, A. *Field Guide to the Water Life of Britain* Reader's Digest Association, London, 1984 (reprinted 1994).

Earll, R.C. *The SEASEARCH Habitat Guide – an identification guide to the main habitats found in the shallow seas around the British Isles* Marine Conservation Society, Ross-on-Wye, 1992.

Erwin, D. & Picton, B. *Guide to Inshore Marine Life* Immel, London, 1987.

Fischer, W. (ed.) *FAO species identification sheets for fishery purposes – Mediterranean and Black Sea (Fishing Area 37)* Vols 1 and 2, FAO, Rome, 1973.

Fischer, W., Bianchi, G., & Scott, W.B. (eds) *FAO species identification sheets for fishery purposes – Eastern Central Atlantic (Fishing Areas 34 and 47 – in part)* Vols 1-6, Department of Fisheries and Oceans, Ottawa, 1981.

Gosner, K.L. *A Field Guide to the Atlantic Seashore, from the Bay of Fundy to Cape Hatteras* The Peterson FieldGuide series, No.24, Houghton Mifflin, Boston, 1978.

George, J.D. & George, J.J. *Marine Life* Lionel Leventhal, London, 1979.

Graham, A. *Molluscs: prosobranch and pyramellid Gastropods* No. 2 in *Synopses of the British Fauna. New Series* Brill/Backhuys in association with The Linnean Society of London and The Estuarine and Brackish-water Sciences Association, 1988.

Gubbay, S. *A Coastal Directory for Marine Nature Conservation* Marine Conservation Society, Ross-on-Wye, 1988.

Hayward, P.J. & Ryland, J.S. (eds) *Handbook of the Marine Fauna of North-West Europe* Oxford University Press, Oxford, 1995.

Ingle, R.W. *Shallow-water Crabs* No. 25 in *Synopses of the British Fauna. New Series* Cambridge University Press in association with The Linnean Society of London, 1983.

Kozloff, E.N. *Marine Invertebrates of the Pacific Northwest* University of Washington Press, Seattle and London, 1987.

Kozloff, E.N. *Seashore Life of the Northern Pacific Coast* University of Washington Press, Seattle and London, 1983.

Lythgoe, J. & Lythgoe, G. *Fishes of the Sea* Blandford, London, 1991.

MacGarvin, M. *The North Sea* in *Greenpeace: The Seas of Europe* series, Collins & Brown, London, 1990.

Manuel, R.L. *The Anthozoa of the British Isles: a colour guide* Marine Conservation Society, Ross-on-Wye, second edition 1983.

Palmer, R. *Undersea Britain* Immel, London, 1990.

Picton, B. *Ascidians of the British Isles: a colour guide* Marine Conservation Society, Ross-on-Wye, 1985.

Picton, B. *A Field Guide to the Shallow-water Echinoderms of the British Isles* Immel, London, 1993.

Picton, B., & Morrow, C.C. *A Field Guide to the Nudibranchs of the British Isles* Immel, London, 1994

Pitkin, L. *The Living Sea: A Photographic Exploration of Life in the Sea* Fountain Press, Kingston-upon-Thames, 1995.

Ruppert, E.E., & Fox, R.S. *Seashore Animals of the Southeast: a guide to common shallow-water invertebrates of the southeeastern Atlantic Coast* University of South Carolina Press, Columbia, 1988.

Smith, R.I., & Carlton, J.T. (eds) *Light's Manual: Intertidal Invertebrates of the Central California Coast* University of California Press, Berkeley, Los Angeles and London, third edition, 1975.

Thompson, T.E. *Biology of Opisthobranch Molluscs* Vol. 1 Ray Society, London, 1976.

Thompson, T.E. & Brown, G.H. *Biology of Opisthobranch Molluscs* Vol. 2, Ray Society, London, 1984.

Wheeler, A. *Key to the Fishes of Northern Europe* Warne, London, 1978.

Wheeler, A. *The World Encyclopedia of Fishes* Macdonald, London, 1985 edition.

Wheeler, A. *The Pocket Guide to Saltwater Fishes of Britain and Europe* Dragon's World, London, 1992.

Whitehead, P.J.P., Bauchot, M.-L., Hureau, J.-C., Nielsen, J. and Tortonese, E. (eds) *Fishes of the North-eastern Atlantic and the Mediterranean* Vols 1-3, UNESCO, Paris, 1984-1986.

Wood, E. *Subtidal Ecology* Edward Arnold, London, 1987.

Wood, E. *Sea Life of Britain & Ireland* Immel, London, 1988.

Young, A. *Marine Wildlife of Atlantic Europe* Immel, London, 1994.

*The small hermit crab* (Pagurus bernhardus) *walking among tufts of seaweed on a shallow slope.*

# INDEX

*Page numbers in italic indicate that the reference appears in a picture caption.*